Petroleum Exploration: Techniques and Technologies

Petroleum Exploration: Techniques and Technologies

Edited by
Chase Jones

Larsen & Keller
www.larsen-keller.com

Petroleum Exploration: Techniques and Technologies
Edited by Chase Jones
ISBN: 978-1-63549-216-3 (Hardback)

© 2017 Larsen & Keller

Larsen & Keller

Published by Larsen and Keller Education,
5 Penn Plaza,
19th Floor,
New York, NY 10001, USA

Cataloging-in-Publication Data

Petroleum exploration : techniques and technologies / edited by Chase Jones.
    p. cm.
Includes bibliographical references and index.
ISBN 978-1-63549-216-3
1. Petroleum--Prospecting. 2. Petroleum engineering. 3. Oil well logging. 4. Oil wells.
I. Jones, Chase.
TN271.P4 P48 2017
622.1828--dc23

The publisher's policy is to use permanent paper from mills that operate a sustainable forestry policy. Furthermore, the publisher ensures that the text paper and cover boards used have met acceptable environmental accreditation standards.

Printed and bound in the United States of America.

For more information regarding Larsen and Keller Education and its products, please visit the publisher's website www.larsen-keller.com

Table of Contents

Permissions

Index

Preface

Petroleum exploration or hydrocarbon exploration is the science of exploring oil and gas from under the Earth's surface. It is practiced by petroleum geologists and geophysicists. Petroleum exploration combines physical laws with chemistry and computational modeling for surveying, identification and evaluation. This book elucidates the concepts and innovative models around prospective developments with respect to petroleum exploration. It picks up individual branches and explains their need and contribution in the context of the growth of this field. Most of the topics introduced in this text cover new techniques and the applications of petroleum exploration. Different approaches, evaluations and methodologies have been included in it. This textbook, with its detailed analyses and data, will prove immensely beneficial to students involved in this area at various levels.

To facilitate a deeper understanding of the contents of this book a short introduction of every chapter is written below:

Chapter 1- Petroleum exploration or hydrocarbon exploration refers to the search for oil and gas deposits beneath the Earth's surface. Though this was a laborious process some decades prior to the advancement in technology, petroleum exploration has become easier and the success rate of every prospective find has increased substantially. This chapter provides a quick overview of petroleum exploration.

Chapter 2- The chapter strategically encompasses and incorporates the major components and key concepts of petroleum exploration, providing a complete understanding. Every exploration begins with a prospect which is then studied in detail to understand its viability. The reader is introduced to terminology like bright spot, flat spot, dim spot and polarity reversal. The chapter also provides the reader with a list of oil fields and has a section about topics like abiogenic petroleum origin, global strategic petroleum reserves, petroleum seep, petroleum play etc.

Chapter 3- Once a prospect is identified and evaluated, an exploration well is drilled to determine the presence or absence of oil or gas. Exploration geophysics, seismic refraction, gravity gradiometry, electrical resistivity tomography, aeromagnetic survey, magnetotellurics, ground-penetrating radar and transient electromagnetics are the various exploration techniques discussed in the chapter.

Chapter 4- Well logging records formations under the earth that have been made in the process of digging a borehole. This chapter focuses on the various methods and techniques used in well logging like density logging, formation evaluation neutron porosity, sonic logging, gamma ray logging, spontaneous potential logging and mud logging. The chapter gives detailed information about the scope and service provided by each of the methods.

Chapter 5- An oil well is usually drilled to access deposits of oil and gas that are available below the surface of the Earth. Sometimes there can be mishaps in the drilling and extraction of hydrocarbons from the oil well. In order to avoid such contingencies from occurring, measures have to be taken. This chapter explores the facets of oil well control, hydraulic fracturing and offshore drilling and their fundamental terminology. There is a section on oil platform as well.

Chapter 6- The environmental impact of petroleum exploration is often adverse and affects land and water use, causes noise pollution, water contamination by oil spills and other related problems. This chapter explores the negative impact petroleum exploration has and provides a comprehensive analysis of each detrimental effect. The chapter also provides insightful information on the Hubbert peak theory, Hirsch report and mitigation of peak oil.

Finally, I would like to thank the entire team involved since the inception of this book for their valuable time and contribution. This book would not have been possible without their efforts. I would also like to thank my friends and family for their constant support.

Editor

Introduction to Petroleum Exploration

Petroleum exploration or hydrocarbon exploration refers to the search for oil and gas deposits beneath the Earth's surface. Though this was a laborious process some decades prior to the advancement in technology, petroleum exploration has become easier and the success rate of every prospective find has increased substantially. This chapter provides a quick overview of petroleum exploration.

Hydrocarbon exploration (or oil and gas exploration) is the search by petroleum geologists and geophysicists for hydrocarbon deposits beneath the Earth's surface, such as oil and natural gas. Oil and gas exploration are grouped under the science of petroleum geology.

Onshore Drilling Rig

Exploration Methods

Visible surface features such as oil seeps, natural gas seeps, pockmarks (underwater craters caused by escaping gas) provide basic evidence of hydrocarbon generation (be it shallow or deep in the Earth). However, most exploration depends on highly sophisticated technology to detect and determine the extent of these deposits using exploration geophysics. Areas thought to contain hydrocarbons are initially subjected to a gravity survey, magnetic survey, passive seismic or regional seismic reflection surveys to detect large-scale features of the sub-surface geology. Features of interest (known as *leads*) are subjected to more detailed seismic surveys which work on the principle of the time it takes for reflected sound waves to travel through matter (rock) of varying densities and using the process of depth conversion to create a profile of the substructure. Finally, when a prospect has been identified and evaluated and passes the oil company's selection criteria, an exploration well is drilled in an attempt to conclusively determine the presence or absence of oil or gas.

Oil exploration is an expensive, high-risk operation. Offshore and remote area exploration is generally only undertaken by very large corporations or national governments. Typical shallow shelf oil wells (e.g. North Sea) cost US$10 – 30 million, while deep water wells can cost up to US$100 million plus. Hundreds of smaller companies search for onshore hydrocarbon deposits worldwide, with some wells costing as little as US$100,000.

Elements of a Petroleum Prospect

Mud log in process, a common way to study the rock types when drilling oil wells.

A prospect is a potential trap which geologists believe may contain hydrocarbons. A significant amount of geological, structural and seismic investigation must first be completed to redefine the potential hydrocarbon drill location from a lead to a prospect. Four geological factors have to be present for a prospect to work and if any of them fail neither oil nor gas will be present.

- A source rock - When organic-rich rock such as oil shale or coal is subjected to high pressure and temperature over an extended period of time, hydrocarbons form.

- Migration - The hydrocarbons are expelled from source rock by three density-related mechanisms: the newly matured hydrocarbons are less dense than their precursors, which causes over-pressure; the hydrocarbons are lighter, and so migrate upwards due to buoyancy, and the fluids expand as further burial causes increased heating. Most hydrocarbons migrate to the surface as oil seeps, but some will get trapped.

- Reservoir - The hydrocarbons are contained in a reservoir rock. This is commonly a porous sandstone or limestone. The oil collects in the pores within the rock although open fractures within non-porous rocks (e.g. fractured granite) may also store hydrocarbons. The reservoir must also be permeable so that the hydrocarbons will flow to surface during production.

- Trap - The hydrocarbons are buoyant and have to be trapped within a structural (e.g. Anticline, fault block) or stratigraphic trap. The hydrocarbon trap has to be covered by an impermeable rock known as a seal or cap-rock in order to prevent hydrocarbons escaping to the surface

Exploration Risk

Hydrocarbon exploration is a high risk investment and risk assessment is paramount for successful project portfolio management. Exploration risk is a difficult concept and is usually defined by assigning confidence to the presence of the imperative geological factors, as discussed above. This confidence is based on data and/or models and is usually mapped on Common Risk Segment Maps (CRS Maps). High confidence in the presence of imperative geological factors is usually coloured green and low confidence coloured red. Therefore, these maps are also called Traffic Light Maps, while the full procedure is often referred to as Play Fairway Analysis. The aim of such procedures is to force the geologist to objectively assess all different geological factors. Furthermore, it results in simple maps that can be understood by non-geologists and managers to base exploration decisions on.

Terms used in Petroleum Evaluation

- Bright spot - On a seismic section, coda that have high amplitudes due to a formation containing hydrocarbons.

- Chance of success - An estimate of the chance of all the elements within a prospect working, described as a probability.

- Dry hole - A boring that does not contain commercial hydrocarbons.

- Flat spot - Possibly an oil-water, gas-water or gas-oil contact on a seismic section; flat due to gravity.

- Hydrocarbon in place - amount of hydrocarbon likely to be contained in the prospect. This is calculated using the volumetric equation - GRV x N/G x Porosity x Sh / FVF

    o GRV - Gross rock volume - amount of rock in the trap above the hydrocarbon water contact

    o N/G - net/gross ratio - proportion of the GRV formed by the reservoir rock (range is 0 to 1)

    o Porosity - percentage of the net reservoir rock occupied by pores (typically 5-35%)

    o Sh - hydrocarbon saturation - some of the pore space is filled with water - this must be discounted o   FVF - formation volume factor - oil shrinks and gas expands when brought to the surface. The FVF converts volumes at reservoir conditions (high pressure and high temperature) to storage and sale conditions

- Lead - Potential accumulation is currently poorly defined and requires more data acquisition and/or evaluation in order to be classified as a prospect.

- Play - An area in which hydrocarbon accumulations or prospects of a given type occur. For example, the shale gas plays in North America include the Barnett, Eagle Ford, Fayetteville, Haynesville, Marcellus, and Woodford, among many others.

- Prospect - a lead which has been more fully evaluated.

- Recoverable hydrocarbons - amount of hydrocarbon likely to be recovered during production. This is typically 10-50% in an oil field and 50-80% in a gas field.

Licensing

Petroleum resources are typically owned by the government of the host country. In the USA most onshore (land) oil and gas rights (OGM) are owned by private individuals, in which case oil companies must negotiate terms for a lease of these rights with the individual who owns the OGM. Sometimes this is not the same person who owns the land surface. In most nations the government issues licences to explore, develop and produce its oil and gas resources, which are typically administered by the oil ministry. There are several different types of licence. Oil companies often operate in joint ventures to spread the risk; one of the companies in the partnership is designated the operator who actually supervises the work.

- Tax and Royalty - Companies would pay a royalty on any oil produced, together with a profits tax (which can have expenditure offset against it). In some cases there are also various bonuses and ground rents (license fees) payable to the government - for example a signature bonus payable at the start of the licence. Licences are awarded in competitive bid rounds on the basis of either the size of the work programme (number of wells, seismic etc.) or size of the signature bonus.

- Production Sharing contract (PSA) - A PSA is more complex than a Tax/Royalty system - The companies bid on the percentage of the production that the host government receives (this may be variable with the oil price), There is often also participation by the Government owned National Oil Company (NOC). There are also various bonuses to be paid. Development expenditure is offset against production revenue.

- Service contract - This is when an oil company acts as a contractor for the host government, being paid to produce the hydrocarbons.

Reserves and Resources

Resources are hydrocarbons which may or may not be produced in the future. A resource number may be assigned to an undrilled prospect or an unappraised discovery. Appraisal by drilling additional delineation wells or acquiring extra seismic data will confirm the size of the field and lead to project sanction. At this point the relevant government body gives the oil company a production licence which enables the field to be developed. This is also the point at which oil reserves and gas reserves can be formally booked.

Oil and Gas Reserves

Oil and gas reserves are defined as volumes that will be commercially recovered in the future. Reserves are separated into three categories: proved, probable, and possible. To be included in any reserves category, all commercial aspects must have been addressed, which includes government consent. Technical issues alone separate proved from unproved categories. All reserve estimates involve some degree of uncertainty.

- Proved reserves are the highest valued category. Proved reserves have a "reasonable certainty" of being recovered, which means a high degree of confidence that the volumes will be recovered. Some industry specialists refer to this as P90, i.e., having a 90% certainty of being produced. The SEC provides a more detailed definition:

Proved oil and gas reserves are those quantities of oil and gas, which, by analysis of geoscience and engineering data, can be estimated with reasonable certainty to be economically producible—from a given date forward, from known reservoirs, and under existing economic conditions, operating methods, and government regulations—prior to the time at which contracts providing the right to operate expire, unless evidence indicates that renewal is reasonably certain, regardless of whether deterministic or probabilistic methods are used for the estimation. The project to extract the hydrocarbons must have commenced or the operator must be reasonably certain that it will commence the project within a reasonable time.

- Probable reserves are volumes defined as "less likely to be recovered than proved, but more certain to be recovered than Possible Reserves". Some industry specialists refer to this as P50, i.e., having a 50% certainty of being produced.
- Possible reserves are reserves which analysis of geological and engineering data suggests are less likely to be recoverable than probable reserves. Some industry specialists refer to this as P10, i.e., having a 10% certainty of being produced.

The term 1P is frequently used to denote proved reserves; 2P is the sum of proved and probable reserves; and 3P the sum of proved, probable, and possible reserves. The best estimate of recovery from committed projects is generally considered to be the 2P sum of proved and probable reserves. Note that these volumes only refer to currently justified projects or those projects already in development.

Reserve Booking

Oil and gas reserves are the main asset of an oil company. Booking is the process by which they are added to the balance sheet.

In the United States, booking is done according to a set of rules developed by the Society of Petroleum Engineers (SPE). The reserves of any company listed on the New York Stock Exchange have to be stated to the U.S. Securities and Exchange Commission. Reported reserves may be audited by outside geologists, although this is not a legal requirement.

In Russia, companies report their reserves to the State Commission on Mineral Reserves (GKZ).

Essential Concepts of Petroleum Exploration

The chapter strategically encompasses and incorporates the major components and key concepts of petroleum exploration, providing a complete understanding. Every exploration begins with a prospect which is then studied in detail to understand its viability. The reader is introduced to terminology like bright spot, flat spot, dim spot and polarity reversal. The chapter also provides the reader with a list of oil fields and has a section about topics like abiogenic petroleum origin, global strategic petroleum reserves, petroleum seep, petroleum play, etc.

Petroleum Reservoir

Screenshot of a structure map generated by Contour map software for an 8500ft deep gas & oil reservoir in the Erath field, Erath, Louisiana. The left-to-right gap, near the top of the contour map indicates a Fault line. This fault line is between the blue/green contour lines and the purple/red/yellow contour lines. The thin red circular contour line in the middle of the map indicates the top of the oil reservoir. Because gas floats above oil, the thin red contour line marks the gas/oil contact zone.

A petroleum reservoir or oil and gas reservoir is a subsurface pool of hydrocarbons contained in porous or fractured rock formations. Petroleum reservoirs are broadly classified as conventional and unconventional reservoirs. In case of conventional reservoirs, the naturally occurring hydrocarbons, such as crude oil or natural gas, are trapped by overlying rock formations with lower permeability. While in unconventional reservoirs the rocks have high porosity and low permeability which keeps the hydrocarbons trapped in place, therefore not requiring a cap rock. Reservoirs are found using hydrocarbon exploration methods.

Formation

Crude oil found in all oil reservoirs formed in the Earth's crust from the remains of once-living things. Crude oil is properly known as petroleum, and is used as fossil fuel. Evidence indicates that millions of years of heat and pressure changed the remains of microscopic plant and animal into oil and natural gas.

Roy Nurmi, an interpretation adviser for Schlumberger, described the process as follows: "Plankton and algae, proteins and the life that's floating in the sea, as it dies, falls to the bottom, and these organisms are going to be the source of our oil and gas. When they're buried with the accumulating sediment and reach an adequate temperature, something above 50 to 70 °C they start to cook. This transformation, this change, changes them into the liquid hydrocarbons that move and migrate, will become our oil and gas reservoir."

In addition to the aquatic environment, which is usually a sea, but might also be a river, lake, coral reef or algal mat, the formation of an oil or gas reservoir also requires a sedimentary basin that passes through four steps: deep burial under sand and mud, pressure cooking, hydrocarbon migration from the source to the reservoir rock, and trapping by impermeable rock. Timing is also an important consideration; it is suggested that the Ohio River Valley could have had as much oil as the Middle East at one time, but that it escaped due to a lack of traps. The North Sea, on the other hand, endured millions of years of sea level changes that successfully resulted in the formation of more than 150 oilfields.

Although the process is generally the same, various environmental factors lead to the creation of a wide variety of reservoirs. Reservoirs exist anywhere from the land surface to 30,000 ft (9,000 m) below the surface and are a variety of shapes, sizes and ages.

Traps

A trap forms when the buoyancy forces driving the upward migration of hydrocarbons through a permeable rock cannot overcome the capillary forces of a sealing medium. The timing of trap formation relative to that of petroleum generation and migration is crucial to ensuring a reservoir can form.

Petroleum geologists broadly classify traps into three categories that are based on their geological characteristics: the structural trap, the stratigraphic trap and the far less common hydrodynamic trap. The trapping mechanisms for many petroleum reservoirs have characteristics from several categories and can be known as a combination trap.

Structural Traps

Fold (structural) trap

Fault (structural) trap

Structural traps are formed as a result of changes in the structure of the subsurface due to processes such as folding and faulting, leading to the formation of domes, anticlines, and folds. Examples of this kind of trap are an anticline trap, a fault trap and a salt dome trap.

They are more easily delineated and more prospective than their stratigraphic counterparts, with the majority of the world's petroleum reserves being found in structural traps.

Stratigraphic Traps

Stratigraphic traps are formed as a result of lateral and vertical variations in the thickness, texture, porosity or lithology of the reservoir rock. Examples of this type of trap are an unconformity trap, a lens trap and a reef trap.

Hydrodynamic Traps

Hydrodynamic traps are a far less common type of trap. They are caused by the differences in water pressure, that are associated with water flow, creating a tilt of the hydrocarbon-water contact.

Seals

The seal is a fundamental part of the trap that prevents hydrocarbons from further upward migration.

A capillary seal is formed when the capillary pressure across the pore throats is greater than or equal to the buoyancy pressure of the migrating hydrocarbons. They do not allow fluids to migrate across them until their integrity is disrupted, causing them to leak. There are two types of capillary seal whose classifications are based on the preferential mechanism of leaking: the hydraulic seal and the membrane seal.

The membrane seal will leak whenever the pressure differential across the seal exceeds the threshold displacement pressure, allowing fluids to migrate through the pore spaces in the seal. It will leak just enough to bring the pressure differential below that of the displacement pressure and will reseal.

The hydraulic seal occurs in rocks that have a significantly higher displacement pressure such that the pressure required for tension fracturing is actually lower than the pressure required for fluid displacement – for example, in evaporites or very tight shales. The rock will fracture when the pore pressure is greater than both its minimum stress and its tensile strength then reseal when the pressure reduces and the fractures close.

Estimating Reserves

After the discovery of a reservoir, a petroleum engineer will seek to build a better picture of the accumulation. In a simple textbook example of a uniform reservoir, the first stage is to conduct a seismic survey to determine the possible size of the trap. Appraisal wells can be used to determine the location of oil-water contact and with it, the height of the oil bearing sands. Often coupled with seismic data, it is possible to estimate the volume of oil bearing reservoir.

The next step is to use information from appraisal wells to estimate the porosity of the rock. The porosity, or the percentage of the total volume that contains fluids rather than solid rock, is 20-35% or less. It can give information on the actual capacity. Laboratory testing can determine the characteristics of the reservoir fluids, particularly the expansion factor of the oil, or how much the oil expands when brought from high pressure, high temperature of the reservoir to "stock tank" at the surface.

With such information, it is possible to estimate how many "stock tank" barrels of oil are located in the reservoir. Such oil is called the stock tank oil initially in place (STOIIP). As a result of studying factors such as the permeability of the rock (how easily fluids can flow through the rock) and possible drive mechanisms, it is possible to estimate the recovery factor, or what proportion of oil in place can be reasonably expected to be produced. The recovery factor is commonly 30-35%, giving a value for the recoverable reserves.

The difficulty is that reservoirs are not uniform. They have variable porosities and permeabilities and may be compartmentalised, with fractures and faults breaking them up and complicating fluid flow. For this reason, computer modeling of economically viable reservoirs is often carried out. Geologists, geophysicists and reservoir engineers work together to build a model which allows simulation of the flow of fluids in the reservoir, leading to an improved estimate of reserves.

Production

To obtain the contents of the oil reservoir, it is usually necessary to drill into the Earth's crust, although surface oil seeps exist in some parts of the world, such as the La Brea tar pits in California, and numerous seeps in Trinidad.

Drive Mechanisms

A virgin reservoir may be under sufficient pressure to push hydrocarbons to surface. As the fluids are produced, the pressure will often decline, and production will falter. The reservoir may respond to the withdrawal of fluid in a way that tends to maintain the pressure. Artificial drive methods may be necessary.

Solution Gas Drive

This mechanism (also known as depletion drive) depends on the associated gas of the oil. The virgin reservoir may be entirely liquid, but will be expected to have gaseous hydrocarbons in solution due to the pressure. As the reservoir depletes, the pressure falls below the bubble point, and the gas comes out of solution to form a gas cap at the top. This gas cap pushes down on the liquid helping to maintain pressure.

This occurs when the natural gas is in a cap below the oil. When the well is drilled the lowered pressure above means that the oil expands. As the pressure is reduced it reaches bubble point and subsequently the gas bubbles drive the oil to the surface. The bubbles then reach critical saturation and flow together as a single gas phase. Beyond this point and below this pressure the gas phase flows out more rapidly than the oil because of its lowered viscosity. More free gas is produced and eventually the energy source is depleted. In some cases depending on the geology the gas may migrate to the top of the oil and form a secondary gas cap.

Some energy may be supplied by water, gas in water, or compressed rock. These are usually minor contributions with respect to hydrocarbon expansion.

By properly managing the production rates, greater benefits can be had from solution gas drives. Secondary recovery involves the injection of gas or water to maintain reservoir pressure. The gas/oil ratio and the oil production rate are stable until the reservoir pressure drops below the bubble point when critical gas saturation is reached. When the gas is exhausted, the gas/oil ratio and the oil rate drops, the reservoir pressure has been reduced and the reservoir energy exhausted.

Gas Cap Drive

In reservoirs already having a gas cap (the virgin pressure is already below bubble point), the gas cap expands with the depletion of the reservoir, pushing down on the liquid sections applying extra pressure.

This is present in the reservoir if there is more gas than can be dissolved in the reservoir. The gas will often migrate to the crest of the structure. It is compressed on top of the oil reserve, as the oil is produced the cap helps to push the oil out. Over time the gas cap moves down and infiltrates the oil and eventually the well will begin to produce more and more gas until it produces only gas. It is best to manage the gas cap effectively; that is, placing the oil wells such that the gas cap will not reach them until the maximum amount of oil is produced. Also a high production rate may cause the gas to migrate downward into the production interval. In this case over time the reservoir pressure depletion is not as steep as in the case of solution based gas drive. In this case the oil rate will not decline as steeply but will depend also on the placement of the well with respect to the gas cap.

As with other drive mechanisms, water or gas injection can be used to maintain reservoir pressure. When a gas cap is coupled with water influx the recovery mechanism can be highly efficient.

Aquifer (Water) Drive

Water (usually salty) may be present below the hydrocarbons. Water, as with all liquids, is compressible to a small degree. As the hydrocarbons are depleted, the reduction in pressure in the reservoir allows the water to expand slightly. Although this unit expansion is minute, if the aquifer is large enough this will translate into a large increase in volume, which will push up on the hydrocarbons, maintaining pressure.

With a water-drive reservoir the decline in reservoir pressure is very slight; in some cases the reservoir pressure may remain unchanged. The gas/oil ratio also remains stable. The oil rate will remain fairly stable until the water reaches the well. In time, the water cut will increase and the well will be watered out.

The water may be present in an aquifer (but rarely one replenished with surface water). This water gradually replaces the volume of oil and gas that is produced out of the well, given that the production rate is equivalent to the aquifer activity. That is, the aquifer is being replenished from some natural water influx. If the water begins to be produced along with the oil, the recovery rate may become uneconomical owing to the higher lifting and water disposal costs.

Water and Gas Injection

If the natural drives are insufficient, as they very often are, then the pressure can be artificially maintained by injecting water into the aquifer or gas into the gas cap.

Gravity Drainage

The force of gravity will cause the oil to move downward of the gas and upward of the water. If vertical permeability exists then recovery rates may be even better.

Gas & Gas Condensate Reservoirs

These occur if the reservoir conditions allow the hydrocarbons to exist as a gas. Retrieval is a matter of gas expansion. Recovery from a closed reservoir (i.e., no water drive) is very good, especially if bottom hole pressure is reduced to a minimum (usually done with compressors at the well head). Any produced liquids are light coloured to colourless, with a gravity higher than 45 API.

Gas Cycling is the process where dry gas is injected and produced along with condensed liquid.

Oil Reserves

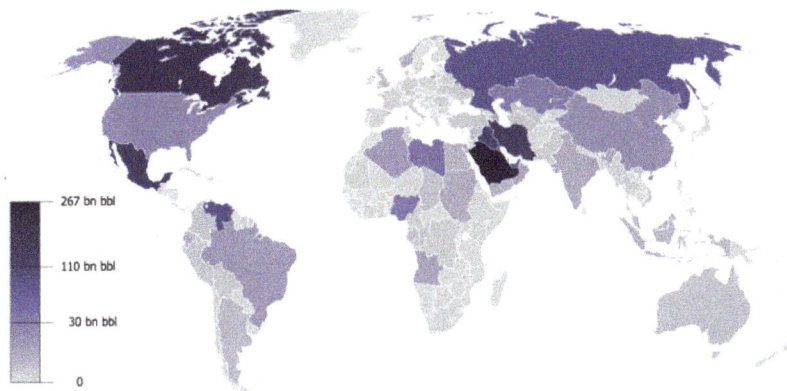

A map of world oil reserves, 2013.

Oil reserves are the amount of technically and economically recoverable oil. Reserves may be for a well, for a reservoir, for a field, for a nation, or for the world. Different classifications of reserves are related to their degree of certainty.

The total estimated amount of oil in an oil reservoir, including both producible and non-producible oil, is called *oil in place*. However, because of reservoir characteristics and limitations in petroleum extraction technologies, only a fraction of this oil can be brought to the surface, and it is only this producible fraction that is considered to be *reserves*. The ratio of reserves to the total amount of oil in a particular reservoir is called the *recovery factor*. Determining a recovery factor for a given field depends on several features of the operation, including method of oil recovery used and technological developments.

Based on data from OPEC at the beginning of 2013 the highest proved oil reserves including non-conventional oil deposits are in Venezuela (20% of global reserves), Saudi Arabia (18% of global reserves), Canada (13% of global reserves), and Iran (9%).

Because the geology of the subsurface cannot be examined directly, indirect techniques must be used to estimate the size and recoverability of the resource. While new technologies have increased the accuracy of these techniques, significant uncertainties still remain. In general, most early estimates of the reserves of an oil field are conservative and tend to grow with time. This phenomenon is called *reserves growth*.

Many oil-producing nations do not reveal their reservoir engineering field data and instead provide unaudited claims for their oil reserves. The numbers disclosed by some national governments are suspected of being manipulated for political reasons.

Classifications

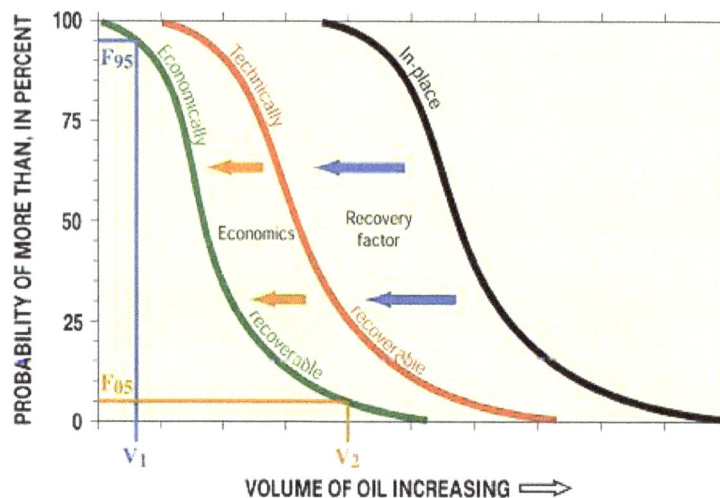

Schematic graph illustrating petroleum volumes and probabilities. Curves represent categories of oil in assessment. There is a 95% chance (i.e., probability, F95) of at least volume V1 of economically recoverable oil, and there is a 5-percent chance (F05) of at least volume V2 of economically recoverable oil.

All reserve estimates involve uncertainty, depending on the amount of reliable geologic and engineering data available and the interpretation of that data. The relative degree of uncertainty can be expressed by dividing reserves into two principal classifications—"proven" (or "proved") and "unproven" (or "unproved"). Unproven reserves can further be divided into two subcategories—"probable" and "possible"—to indicate the relative degree of uncertainty about their existence. The most commonly accepted definitions of these are based on those approved by the Society of Petroleum Engineers (SPE) and the World Petroleum Council (WPC) in 1997.

Proven Reserves

Proven reserves are those reserves claimed to have a *reasonable certainty* (normally at least 90% confidence) of being recoverable under existing economic and political conditions, with existing technology. Industry specialists refer to this as P90 (that is, having a 90% certainty of being produced). Proven reserves are also known in the industry as 1P.

Proven reserves are further subdivided into "proven developed" (PD) and "proven undeveloped" (PUD). PD reserves are reserves that can be produced with existing wells and perforations, or from additional reservoirs where minimal additional investment (operating expense) is required. PUD reserves require additional capital investment (e.g., drilling new wells) to bring the oil to the surface.

Until December 2009 "1P" proven reserves were the only type the U.S. Securities and Exchange Commission allowed oil companies to report to investors. Companies listed on U.S. stock exchanges must substantiate their claims, but many governments and national oil companies do not disclose verifying data to support their claims. Since January 2010 the SEC now allows companies to also provide additional optional information declaring "2P" (both proven and probable) and "3P" (proven + probable + possible) provided the evaluation is verified by qualified third party consultants, though many companies choose to use 2P and 3P estimates only for internal purposes.

Unproven Reserves

An oil well in Canada, which has the world's third largest oil reserves.

Unproven reserves are based on geological and/or engineering data similar to that used in estimates of proven reserves, but technical, contractual, or regulatory uncertainties preclude such reserves being classified as proven. Unproven reserves may be used internally by oil companies and government agencies for future planning purposes but are not routinely compiled. They are sub-classified as *probable* and *possible*.

Probable reserves are attributed to known accumulations and claim a 50% confidence level of recovery. Industry specialists refer to them as "P50" (i.e., having a 50% certainty of being produced). These reserves are also referred to in the industry as "2P" (proven plus probable).

Possible reserves are attributed to known accumulations that have a less likely chance of being recovered than probable reserves. This term is often used for reserves which are claimed to have at least a 10% certainty of being produced ("P10"). Reasons for classifying reserves as possible include varying interpretations of geology, reserves not producible at commercial rates, uncertainty due to reserve infill (seepage from adjacent areas) and projected reserves based on future recovery methods. They are referred to in the industry as "3P" (proven plus probable plus possible).

Russian Reserve Categories

In Russia, reserves categories A, B, and C1 correspond roughly to proved developed producing, proved developed nonproducing, and proved undeveloped, respectively; the designation ABC1 corresponds to proved reserves. The Russian category C2 includes probable and possible reserves.

Strategic Petroleum Reserves

Many countries maintain government-controlled oil reserves for both economic and national security reasons. According to the United States Energy Information Administration, approximately 4.1 billion barrels (650,000,000 m³) of oil are held in strategic reserves, of which 1.4 billion is government-controlled. These reserves are generally not counted when computing a nation's oil reserves.

Resources

Total World Oil Reserves

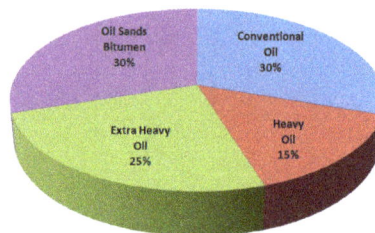

Unconventional oil resources are greater than conventional ones.

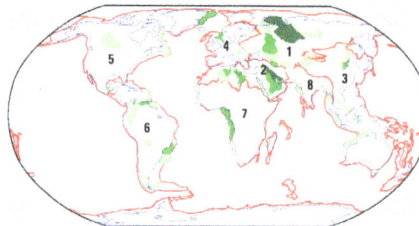

Cumulative oil production plus remaining reserves and undiscovered resources. United States not included.

A more sophisticated system of evaluating petroleum accumulations was adopted in 2007 by the Society of Petroleum Engineers (SPE), World Petroleum Council (WPC), American Association of Petroleum Geologists (AAPG), and Society of Petroleum Evaluation Engineers (SPEE). It incorporates the 1997 definitions for reserves, but adds categories for *contingent resources* and *prospective resources*.

Contingent resources are those quantities of petroleum estimated, as of a given date, to be potentially recoverable from *known* accumulations, but the applied project(s) are not yet considered mature enough for commercial development due to one or more contingencies. Contingent resources may include, for example, projects for which there are no viable markets, or where commercial recovery is dependent on technology under development, or where evaluation of the accumulation is insufficient to clearly assess commerciality.

Prospective resources are those quantities of petroleum estimated, as of a given date, to be potentially recoverable from *undiscovered* accumulations by application of future development projects. Prospective resources have both an associated chance of discovery and a chance of development.

The United States Geological Survey uses the terms *technically* and *economically* recoverable resources when making its petroleum resource assessments. Technically recoverable resources represent that proportion of assessed in-place petroleum that may be recoverable using current recovery technology, without regard to cost. Economically recoverable resources are technically recoverable petroleum for which the costs of discovery, development, production, and transport, including a return to capital, can be recovered at a given market price.

"Unconventional resources" exist in petroleum accumulations that are pervasive throughout a large area. Examples include extra heavy oil, oil sand, and oil shale deposits. Unlike "conventional resources", in which the petroleum is recovered through wellbores and typically requires minimal processing prior to sale, unconventional resources require specialized extraction technology to produce. For example, steam and/or solvents are used to mobilize bitumen for in-situ recovery. Moreover, the extracted petroleum may require significant processing prior to sale (e.g., bitumen upgraders). The total amount of unconventional oil resources in the world considerably exceeds the amount of conventional oil reserves, but are much more difficult and expensive to develop.

Estimation Techniques

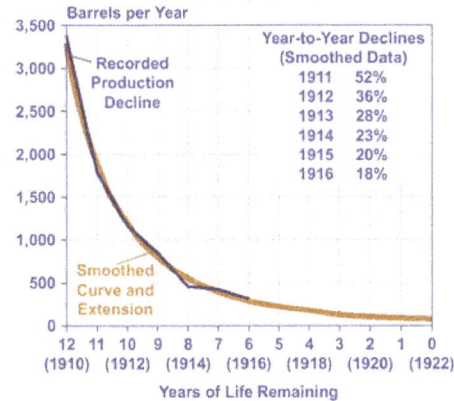

Figure 1. Production Decline Curve for Yearly Production from an Individual Well in the Glenn Pool, 1910-1922

Example of a production decline curve for an individual well

The amount of oil in a subsurface reservoir is called *oil in place* (OIP). Only a fraction of this oil can be recovered from a reservoir. This fraction is called the *recovery factor*. The portion that can be recovered is considered to be a reserve. The portion that is not recoverable is not included unless and until methods are implemented to produce it.

Volumetric Method

Volumetric methods attempt to determine the amount of oil in place by using the size of the reservoir as well as the physical properties of its rocks and fluids. Then a recovery factor is assumed, using assumptions from fields with similar characteristics. OIP is multiplied by the recovery factor to arrive at a reserve number. Current recovery factors for oil fields around the world typically range between 10 and 60 percent; some are over 80 percent. The wide variance is due largely to the diversity of fluid and reservoir characteristics for different deposits. The method is most useful early in the life of the reservoir, before significant production has occurred.

Materials Balance Method

The *materials balance method* for an oil field uses an equation that relates the volume of oil, water and gas that has been produced from a reservoir and the change in reservoir pressure to calculate the remaining oil. It assumes that, as fluids from the reservoir are produced, there will be a change in the reservoir pressure that depends on the remaining volume of oil and gas. The method requires extensive pressure-volume-temperature analysis and an accurate pressure history of the field. It requires some production to occur (typically 5% to 10% of ultimate recovery), unless reliable pressure history can be used from a field with similar rock and fluid characteristics.

Production Decline Curve Method

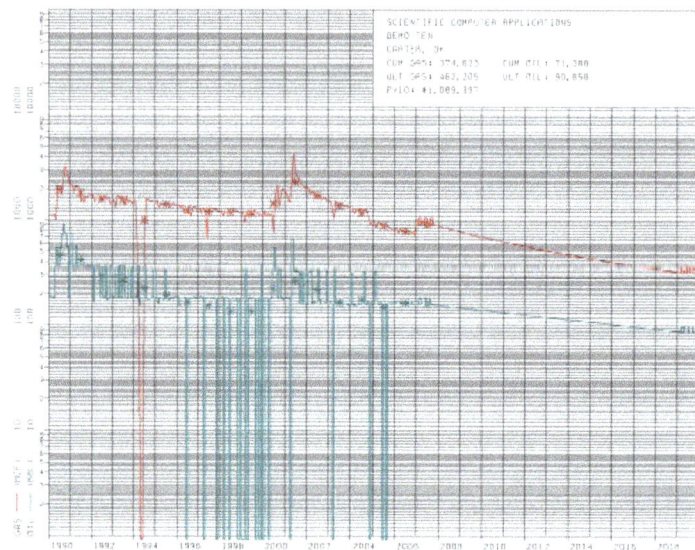

Decline curve generated by decline curve analysis software, utilized in petroleum economics to indicate the depletion of oil & gas in a petroleum reservoir. The Y axis is a semi log scale, indicating the rate of oil depletion (green line), and gas depletion (red line). The X axis is a coordinate scale, indicating time in years and displays the production decline curve. The top red line is the gas decline curve, which is a hyperbolic decline curve. Gas is measured in MCF (thousand cubic feet in this case). The lower Blue line is the oil decline curve, which is an exponential decline curve. Oil is measured in BBL (Oil barrels). Data is from actual sales, not pumped production. The dips to zero indicate there were no sales that month, likely because the oil well did not produce a full tank, and thus was not worth a visit from a tank truck. The upper right legend (map) displays CUM, which is the cumulative gas or oil produced. ULT is the ultimate recovery projected for the well. Pv10 is the discounted present value of 10%, which is the future value of the remaining lease, valued for this oil well at $1.089 million USD.

The *decline curve method* uses production data to fit a decline curve and estimate future oil production. The three most common forms of decline curves are exponential, hyperbolic, and harmonic. It is assumed that the production will decline on a reasonably smooth curve, and so allowances must be made for wells shut in and production restrictions. The curve can be expressed mathematically or plotted on a graph to estimate future production. It has the advantage of (implicitly) including all reservoir characteristics. It requires a sufficient history to establish a statistically significant trend, ideally when production is not curtailed by regulatory or other artificial conditions.

Reserves Growth

Experience shows that initial estimates of the size of newly discovered oil fields are usually too low. As years pass, successive estimates of the ultimate recovery of fields tend to increase. The term *reserve growth* refers to the typical increases in estimated ultimate recovery that occur as oil fields are developed and produced.

Estimated Reserves by Country

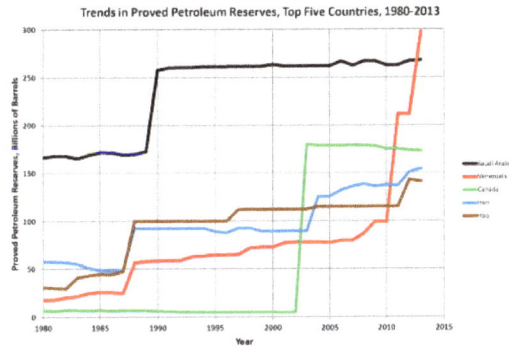

Trends in proved oil reserves in top five countries, 1980-2013 (date from US Energy Information Administration)

BBL = barrel of oil

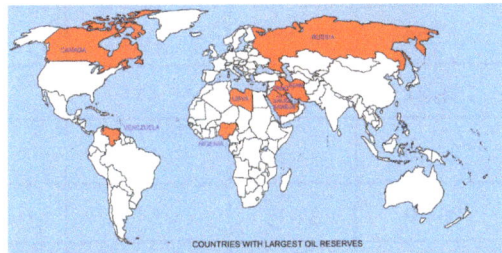

Countries with largest oil reserves

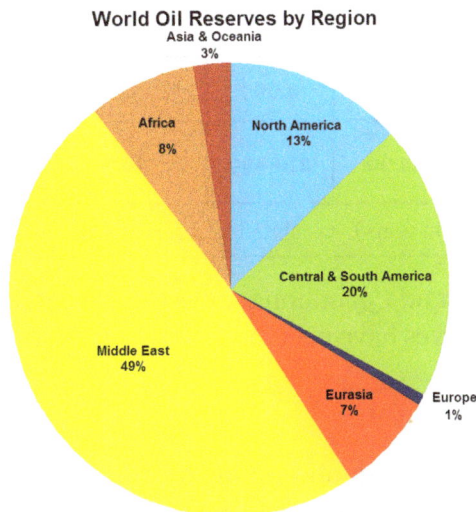

Most of the world's oil reserves are in the Middle East.

Summary of Proven Reserve Data as of 2012						
—	Country	Reserves 10^9 bbl	Reserves 10^9 m³	Production 10^6 bbl/d	Production 10^3 m³/d	Reserve/ Production Ratio[1] years
1	Venezuela	296.50	47.140	2.1	330	387
2	Saudi Arabia	265.40	42.195	8.9	1,410	81
3	Canada	175.00	27.823	2.7	430	178
4	Iran	151.20	24.039	4.1	650	101
5	Iraq	143.10	22.751	3.4	540	163
6	Kuwait	101.50	16.137	2.3	370	121
7	United Arab Emirates	97.80	15.549	2.4	380	156
8	Russia	80.00	12.719	10.0	1,590	22
9	Libya	47.00	7.472	1 7	270	76
10	Nigeria	37.00	5.883	2.5	400	41
11	Kazakhstan	30.00	4.770	1.5	240	55
12	Qatar	25.41	4.040	1.1	170	63
13	China	25.40	4.038	4.1	650	15
14	United States	25.00	3.975	7.0	1,110	10
15	Angola	13.50	2.146	1.9	300	19
16	Algeria	13.42	2.134	1.7	270	22
17	Brazil	13.20	2.099	2.1	330	17
	Total of top seventeen reserves	1,324.00	210.499	56.7	9,010	64

Notes: 1 Reserve to Production ratio (in years), calculated as reserves / annual production. (from above)

It is estimated that between 100 and 135 billion tonnes (which equals between 133 and 180 billions m³ of oil) of the world's oil reserves have been used between 1850 and the present.

OPEC Countries

Since OPEC started to set production quotas on the basis of reserves levels in the 1980s, many of its members have reported significant increases in their official reserves. There are doubts about the reliability of these estimates, which are not provided with any form of verification that meet external reporting standards. The following table illustrates these rises.

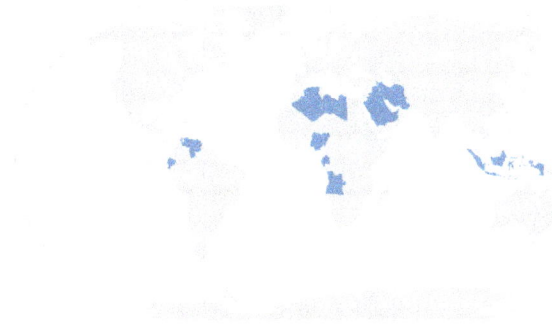

OPEC countries

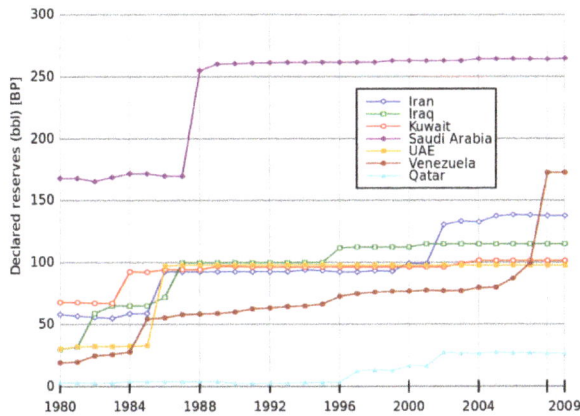

oil reserves of OPEC 1980–2005

Declared reserves of major OPEC Producers (billion of barrels)							
BP Statistical Review - June 2009							
OPEC Annual Statistical Bulletin 2010/2011							
Year	Iran	Iraq	Kuwait	Saudi Arabia	UAE	Venezuela	Libya	Nigeria
1980	58.3	30.0	67.9	168.0	30.4	19.5	20.3	16.7
1981	57.0	32.0	67.7	167.9	32.2	19.9	22.6	16.5
1982	56.1	59.0	67.2	165.5	32.4	24.9	22.2	16.8
1983	55.3	65.0	67.0	168.8	32.3	25.9	21.8	16.6
1984	58.9	65.0	92.7	171.7	32.5	28.0	21.4	16.7
1985	59.0	65.0	92.5	171.5	33.0	54.5	21.3	16.6
1986	92.9	72.0	94.5	169.7	97.2	55.5	22.8	16.1
1987	92.9	100.0	94.5	169.6	98.1	58.1	22.8	16.0
1988	92.9	100.0	94.5	255.0	98.1	58.5	22.8	16.0
1989	92.9	100.0	97.1	260.1	98.1	59.0	22.8	16.0
1990	92.9	100.0	97.0	260.3	98.1	60.1	22.8	17.1
1991	92.9	100.0	96.5	260.9	98.1	62.6	22.8	20.0
1992	92.9	100.0	96.5	261.2	98.1	63.3	22.8	21.0
1993	92.9	100.0	96.5	261.4	98.1	64.4	22.8	21.0
1994	94.3	100.0	96.5	261.4	98.1	64.9	22.8	21.0
1995	93.7	100.0	96.5	261.5	98.1	66.3	29.5	20.8

1996	92.6	112.0	96.5	261.4	97.8	72.7	29.5	20.8
1997	92.6	112.5	96.5	261.5	97.8	74.9	29.5	20.8
1998	93.7	112.5	96.5	261.5	97.8	76.1	29.5	22.5
1999	93.1	112.5	96.5	262.8	97.8	76.8	29.5	29.0
2000	99.5	112.5	96.5	262.8	97.8	76.8	36.0	29.0
2001	99.1	115.0	96.5	262.7	97.8	77.7	36.0	31.5
2002	130.7	115.0	96.5	262.8	97.8	77.3	36.0	34.3
2003	133.3	115.0	99.0	262.7	97.8	77.2	39.1	35.3
2004	132.7	115.0	101.5	264.3	97.8	79.7	39.1	35.9
2005	137.5	115.0	101.5	264.2	97.8	80.0	41.5	36.2
2006	138.4	115.0	101.5	264.3	97.8	87.3	41.5	36.2
2007	138.2	115.0	101.5	264.2	97.8	99.4	43.7	36.2
2008	137.6	115.0	101.5	264.1	97.8	172.3	43.7	36.2
2009	137.0	115.0	101.5	264.6	97.8	211.1	46.4	36.2
2010	151.2	143.1	101.5	264.5	97.8	296.5	47.1	36.2

The sudden revisions in OPEC reserves, totaling nearly 300 bn barrels, have been much debated. Some of it is defended partly by the shift in ownership of reserves away from international oil companies, some of whom were obliged to report reserves under conservative US Securities and Exchange Commission rules. The most prominent explanation of the revisions is prompted by a change in OPEC rules which set production quotas (partly) on reserves. In any event, the revisions in official data had little to do with the actual discovery of new reserves.

Total reserves in many OPEC countries hardly changed in the 1990s. Official reserves in Kuwait, for example, were unchanged at 96.5 Gbbl (15.34×10^9 m³) (including its share of the Neutral Zone) from 1991 to 2002, even though the country produced more than 8 Gbbl (1.3×10^9 m³) and did not make any important new discoveries during that period. The case of Saudi Arabia is also striking, with proven reserves estimated at between 260 and 264 billion barrels (4.20×10^{10} m³) in the past 18 years, a variation of less than 2%, while extracting approximately 60 billion barrels (9.5×10^9 m³) during this period.

Sadad al-Huseini, former head of exploration and production at Saudi Aramco, estimates 300 Gbbl (48×10^9 m³) of the world's 1,200 Gbbl (190×10^9 m³) of proven reserves should be re-categorized as speculative resources, though he did not specify which countries had inflated their reserves. Dr. Ali Samsam Bakhtiari, a former senior expert of the National Iranian Oil Company, has estimated that Iran, Iraq, Kuwait, Saudi Arabia and the United Arab Emirates have overstated reserves by a combined 320–390bn barrels and has said, "As for Iran, the usually accepted official 132 billion barrels (2.10×10^{10} m³) is almost one hundred billion over any realistic assay." *Petroleum Intelligence Weekly* reported that official confidential Kuwaiti documents estimate reserves of Kuwait were only 48 billion barrels (7.6×10^9 m³), of which half were proven and half were possible. The combined value of proven and possible is half of the official public estimate of proven reserves.

In July 2011, OPEC's Annual Statistical Review showed Venezuela's reserves to be larger than Saudi Arabia's.

Prospective Resources

Arctic Prospective Resources

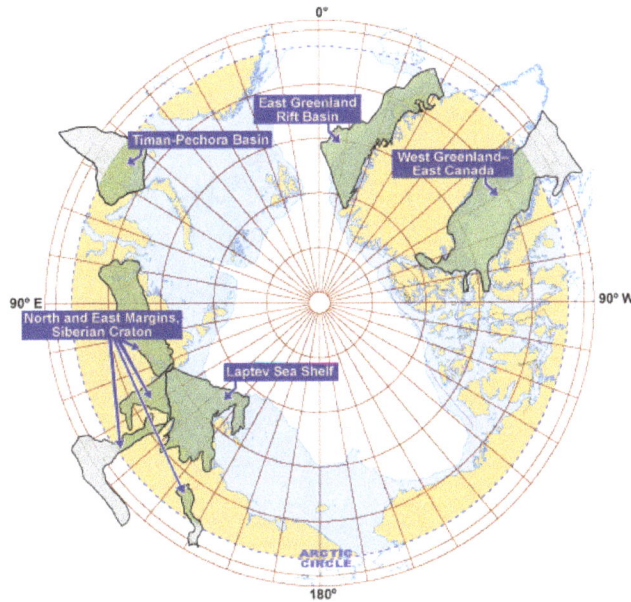

Location of Arctic Basins assessed by the USGS

A 2008 United States Geological Survey estimates that areas north of the Arctic Circle have 90 billion barrels ($1.4×10^{10}$ m^3) of undiscovered, technically recoverable oil and 44 billion barrels ($7.0×10^9$ m^3) of natural gas liquids in 25 geologically defined areas thought to have potential for petroleum. This represented 13% of the expected undiscovered oil in the world. Of the estimated totals, more than half of the undiscovered oil resources were estimated to occur in just three geologic provinces—Arctic Alaska, the Amerasia Basin, and the East Greenland Rift Basins. More than 70% of the mean undiscovered oil resources was estimated to occur in five provinces: Arctic Alaska, Amerasia Basin, East Greenland Rift Basins, East Barents Basins, and West Greenland–East Canada. It was further estimated that approximately 84% of the oil and gas would occur offshore. The USGS did not consider economic factors such as the effects of permanent sea ice or oceanic water depth in its assessment of undiscovered oil and gas resources. This assessment was lower than a 2000 survey, which had included lands south of the Arctic Circle.

Unconventional Prospective Resources

In October 2009, the USGS updated the Orinoco tar sands (Venezuela) value to 513 billion barrels ($8.16×10^{10}$ m^3).

In June 2013 the U.S. Energy Information Administration published a global inventory of estimated recoverable tight oil and tight gas resources in shale formations, "Technically Recoverable Shale Oil and Shale Gas Resources: An Assessment of 137 Shale Formations in 41 Countries Outside the United States." The inventory is incomplete due to exclusion of tight oil and gas from sources other than shale such as sandstone or carbonates, formations underlying the large oil fields located in the Middle East and the Caspian region, off shore formations, or about which there is little information. Estimated technically recoverable shale oil resources total 335 to 345 billion barrels.

List of Oil Fields

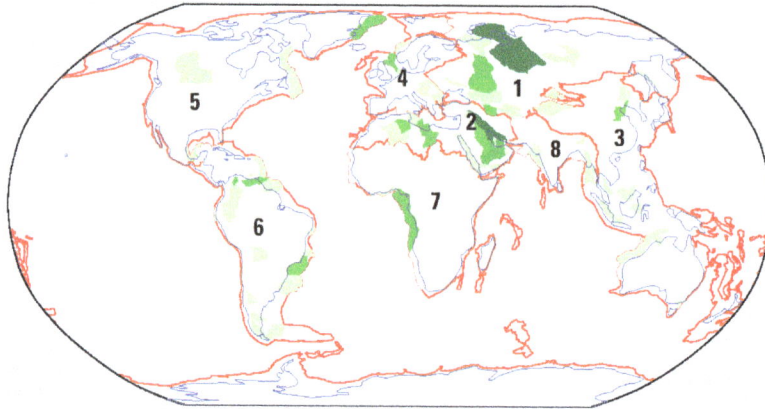

USGS map of countries where oil is located

An oil refinery in Mina-Al-Ahmadi, Kuwait

This list of oil fields includes some major oil fields of the past and present.

The list is incomplete; there are more than 65,000 oil and gas fields of all sizes in the world. However, 94% of known oil is concentrated in fewer than 1500 giant and major fields. Most of the world's largest oilfields are located in the Middle East, but there are also supergiant (>10 billion bbls) oilfields in India, Brazil, Mexico, Venezuela, Kazakhstan, and Russia.

Amounts listed below, in billions of barrels, are the estimated ultimate recoverable petroleum resources (proved reserves plus cumulative production), given historical production and current extraction technology. Oil shale reserves (perhaps 3 trillion barrels (4.8×10^{11} m^3)) and Coal reserves, both of which can be converted to liquid petroleum, are not included in this chart. Other non-conventional liquid fuel sources are similarly excluded in this list.

Oil Fields Greater Than 1 Billion Barrels (160 Million Cubic Metres)

Field	Location	Discovered	Started Production	Peaked	Recoverable Oil, Past and Future (Billion Barrels)	Production (Million Barrels/ day)	Rate of Decline
Ghawar Field	Saudi Arabia	1948	1951	2005, disputed	75-83	5	8% per year
Burgan Field	Kuwait	1937	1948	2005	66-72	1.7	14% per year
Gachsaran Field	Iran	1927	1930	1974	66	0.480	
Ahwaz Field	Iran	1958		1970s	37	.750	Expected to surpass original peak due to new gas injection.
Sugar Loaf field	Brazil	2007			~ 25-40		Not yet developed
Cantarell Field	Mexico	1976	1981	2004	35 18 billion recoverable	.340	peaked in 2004 at 2.14 million barrels per day (340,000 m³/d)
Ku-Maloob-Zaap	Mexico	1979	1981	Production still increasing		.867	Production increasing, most productive Mexican oil field
Bolivar Coastal Field	Venezuela	1917	1922		30-32	2.6-3	
Aghajari Field	Iran	1938	1940		28	0.300	
Azadegan field	Iran	2004			26 (9 recoverable)	0.04	
Lula Field	Brazil, Santos Basin	2007			5-8	0.1	
Safaniya Oil Field	Kuwait/ Saudi Arabia	1951			30	1.2	
Esfandiar Field	Iran				30		
Rumaila Field	Iraq	1953			17	1.3	
Tengiz Field	Kazakhstan	1979	1993	2010	26-40	.53	Expanding from 285k to 1.3 m bpd
Kirkuk Field	Iraq	1927	1934		8.5	0.480	
Shaybah Field	Saudi Arabia				15		
Agha Jari Field	Iran	1937			8.7	.200	
Majnoon Field	Iraq	1975			11-20	0.5	
Samotlor Field	Russia, West Siberia	1965	1969	1980	14-16	0.844	(depletion: 73%) 5% decline per year (2008 - 2014)

Shaikan Sheikh Adi Field	Iraq Kurdistan	2009	2013	Production still increasing	4-6	0.04	Production still increasing
Romashkino Field	Russia Volga-Ural	1948	1949	in decline	16-17	.301 (2006)	depletion: 85%
Prudhoe Bay	United States, Alaska	1967-68	1977	1988	25 (~13 recoverable)	0.66	11% per year
Sarir Field	Libya	1961	1961		12 (6.5 billion recoverable)		
Priobskoye field	Russia, West Siberia	1982	2000		13	0.680 (2008)	14% depleted, Production rapidly expanding.
Lyantorskoye field	Russia, West Siberia	1966	1979		13	0.168 (2004)	depletion: 81%
Abqaiq Field	Saudi Arabia				12	0.43	
Chicontepec Field	Mexico	1926			6.5 (19 certified)		
Berri Field	Saudi Arabia				12		
Zakum Field	Abu Dhabi, UAE	1965	1967		12		
West Qurna Field	Iraq	1973			15-21	0.18-0.25 (pot.)*civil war	
Manifa Field	Saudi Arabia				11		
Fyodorovskoye Field	Russia, West Siberia	1971	1974		11	1.9 (197x)	
East Baghdad Field	Iraq	1976			8	0-0.05 (pot.)*civil war	
Foroozan-Marjan Field	Saudi Arabia/Iran				10		
Marlim Field	Brazil, Campos Basin			in decline	10-14		8% per year
Awali	Bahrain				1		
Azadegan Field	Iran	1999			5.2		
Marun Field	Iran	1963			16	0.52	
Mesopotamian Foredeep Basin	Kuwait				66-72		
Minagish	Kuwait				2		
Raudhatain	Kuwait				11		
Sabriya	Kuwait				3.8-4		
Yibal	Oman		1968		1		

Mukhaizna Oil Field	Oman				1	
Dukhan Field	Qatar		1939		2.2	
Halfaya Field	Iraq				4.1	
Az Zubayr Field	Iraq	1949			6	
Nahr Umr Field	Iraq	1948			6	
Abu-Sa'fah field	Saudi Arabia				6.1	
Hassi Messaoud	Algeria	1956			9	
Bouri Field	Libya	1976	1988		4.5	0.060
Kizomba Complex	Angola				2	
Dalia (oil field)	Angola	1997			1	
Belayim	Angola				>1	
Zafiro	Angola				1	
Zelten oil field	Libya	1956	1961		2.5	
Agbami Field	Nigeria	1998	2008		0.8-1.2	
Bonga Field	Nigeria	1996	2005		1.4	
Azeri-Chirag-Guneshli	Azerbaijan	1985	1997		5.4	0.684
Karachaganak Field	Kazakhstan	1972			2.5	
Kashagan Field	Kazakhstan	2000			30	
Kurmangazy Field	Kazakhstan				6-7	
Darkhan Field	Kazakhstan				9.5	
Zhanazhol Field	Kazakhstan	1960	1987		3	
Uzen Field	Kazakhstan				7	
Kalamkas Field	Kazakhstan				3.2	
Zhetybay Field	Kazakhstan				2.1	
Nursultan Field	Kazakhstan				4.5	
Ekofisk oil field	Norway	1969	1971	2006	3.3	0.127
Troll Vest	Norway	1979	1990	2003	1.4	
Statfjord	Norway	1974	1979	1987	3.4	
Gullfaks	Norway	1978	1986	1994	2.1	
Oseberg	Norway	1979	1988		2.2	
Snorre	Norway	1979	1992	2003	1.5	
Mamontovskoye Field	Russia				8	
Russkoye Field	Russia				2.5	
Kamennoe Field	Russia				1.9	
Vankor Field	Russia	1983	2009		3.8	
Vatyeganskoye Field	Russia				1.4	

Tevlinsko-Russ-kinskoye Field	Russia				1.3		
Sutorminskoye Field	Russia				1.3		
Urengoy group	Russia				1		
Ust-Balykskoe Field	Russia				>1		
Tuymazinskoe Field	Russia				3		
Arlanskoye Field	Russia				>2		
South-Hilchuy Field	Russia				3.1		
North-Dol-ginskoye Field	Russia				2.2		
Nizhne-Chutinskoe Field	Russia				1.7		
South-Dol-ginskoye Field	Russia				1.6		
Prirazlomnoye Field	Russia	1989	2011		1.4		
West-Mat-veevskoye Field	Russia				1.1		
Sakhalin Islands	Russia				14		
Odoptu	Russia				1		
Arukutun-Dagi	Russia				1		
Piltun-Astokhs-koye Field	Russia	1986			1		
Ayash Field East-Odoptu Field	Russia				4.5		
Verhne-Chon-skoye Field	Russia				1.3		
Talakan Field	Russia				1.3		
North-Caucasus Basin	Russia				1.7		
Clair oilfield	United Kingdom	1977			5 (1.75 recoverable)		
Forties oilfield	United Kingdom	1970			5		
Jupiter field	Brazil	2008			7		
Cupiagua/Cu-siana	Colombia				1		
Boscán Field, Venezuela	Venezuela	1946	1947		1.6		
Pembina	Canada	1953	1953				
Swan Hills	Canada						
Rainbow Lake	Canada						

Hibernia	Canada	1979	1997		3		
Terra Nova Field	Canada	1984	2002		1.0		
Kelly-Snyder / SACROC	United States, Texas				1.5		
Bakken Oil Field	United States, North Da-kota	1951			.862		
Yates Oil Field	United States, Texas	1926	1926	1929	3.0 (2.0 billion recovered; 1.0 reserve remaining)		
Kuparuk oil field	United States, Alaska	1969			6		
Alpine	United States, Alaska				0.4-1		
East Texas Oil Field	United States, Texas	1930			6		
Spraberry Trend	United States, Texas	1943			10		
Wilmington Oil Field	United States, California	1932			3		
South Belridge Oil Field	United States, California	1911			2		
Coalinga Oil Field	United States, California	1887			1		
Elk Hills	United States, California	1911			1.5		
Kern River	United States, California	1899			2.5		
Midway-Sunset Field	United States, California	1894			3.4		
Thunder Horse Oil Field	United States, Gulf of Mexico	1999			1	0.25	
Kingfish	Australia				1.2		
Halibut	Australia				1		
Daqing Field	China	1959	1960		16		depletion: 90%

Jidong Field	China				2.2		
Tahe Field	China				8		
Nanpu Oil Field	China				7.35		
Zafiro Field	Equatorial Guinea	1995		2004	1		

Petroleum Play

In geology, a petroleum play, or simply a play, is a group of oil fields or prospects in the same region that are controlled by the same set of geological circumstances. The term is widely and heavily used in the entire realm of exploitation of hydrocarbon-based resources.

The normal steps in the play cycle are:

1. Initial observations of a possible oil reserve

2. Testing and adjustments to initial estimates of extraction

3. High success in locating and extracting oil from a reserve

4. Lower success as the reserve is depleted

5. Continued decrease in further exploration of the region

A particular stratigraphic or structural geologic setting is also often known as a *play*. For example, in a relatively unexplored area such as the Falkland Islands, one might speak of the "Paleozoic play" to refer to the potential oil reserves that might be found within Paleozoic strata. In a well-explored basin such as the Gulf of Mexico, explorationists refer to the "Wilcox play" or the "Norphlet play" to collectively designate the production and possible production from those particular geological formations, of Paleocene and Jurassic age, respectively.

A play may also be a broad category of possible reservoirs or rock types, as in the *turbidite play* of offshore Angola or the carbonate play in the East Java Sea, or to the structural geology of the setting, as in the *sub-upthrust play* of Wyoming. Sometimes the word play is applied to a geographic area with hydrocarbon potential, as the South Texas play or the Niger Delta play, but usually "play" is used with the sense of restricting discussion to exploring a particular geologic setting. Thus, one might have both the Wilcox play and the Norphlet play (among others) in partially overlapping areas of the coast of the Gulf of Mexico; the *Gulf of Mexico deep-water play* might or might not include elements or particular locations appropriate to either the Wilcox or the Norphlet, or both. The term is one of convenience for discussion, and may refer to geologic time intervals, rock types, structures, or some combination of them.

Petroleum Seep

A petroleum seep is a place where natural liquid or gaseous hydrocarbons escape to the earth's atmosphere and surface, normally under low pressure or flow. Seeps generally occur above either terrestrial or offshore petroleum accumulation structures.The hydrocarbons may escape along geological layers, or across them through fractures and fissures in the rock, or directly from an outcrop of oil-bearing rock.

Naturally occurring oil seep near McKittrick, California, United States.

Petroleum seep near the Korňa in northern Slovakia.

Tar "volcano" in the Carpinteria, California asphalt mine. Oil exudes from joint cracks in the petroliferous shale forming the floor of mine. 1906 photo, U.S. Geological Survey Bulletin 321

Petroleum seeps are quite common in many areas of the world, and have been exploited by mankind since paleolithic times. Natural products associated with these seeps include bitumen, pitch, asphalt and tar. In locations where seeps of natural gas are sufficiently large, natural "eternal flames" often persist. The occurrence of surface petroleum was often included in location names that developed; these locations are also associated with early oil and gas exploitation as well as scientific and technological developments, which have grown into the petroleum industry.

History of Petroleum Seep Exploitation

Prehistory

The exploitation of bituminous rocks and natural seep deposits dates back to paleolithic times. The earliest known use of bitumen (natural asphalt) was by Neanderthals some 70,000 years ago, with bitumen adhered to ancient tools found at Neanderthal sites in Syria.

Ancient Civilizations

After the arrival of Homo sapiens, humans used bitumen for construction of buildings and water-proofing of reed boats, among other uses. The use of bitumen for waterproofing and as an adhesive dates at least to the fifth millennium BCE in the early Indus community of Mehrgarh where it was used to line the baskets in which they gathered crops. The material was also used as early as the third millennium BCE in statuary, mortaring brick walls, waterproofing baths and drains, in stair treads, and for shipbuilding. According to Herodotus, and confirmed by Diodorus Siculus, more than four thousand years ago natural asphalt was employed in the construction of the walls and towers of Babylon; there were oil pits near Ardericca (near Babylon), as well as a pitch spring on Zacynthus (Ionian islands, Greece). Great quantities of it were found on the banks of the river Issus, one of the tributaries of the Euphrates.

In ancient times, bitumen was primarily a Mesopotamian commodity used by the Sumerians and Babylonians, although it was also found in the Levant and Persia. Along the Tigris and Euphrates rivers, the area was littered with hundreds of pure bitumen seepages. The Mesopotamians used the bitumen for waterproofing boats and buildings. Ancient Persian tablets indicate the medicinal and lighting uses of petroleum in the upper levels of their society. In ancient Egypt, the use of bitumen was important in creating Egyptian mummies — in fact, the word mummy is derived from the Arab word *mūmiyyah*, which means bitumen. Oil from seeps was exploited in the Roman province of Dacia, now in Romania, where it was called *picula*.

In East Asia these locations were known in China, where the earliest known drilled oil wells date to 347 CE or earlier. The ancient records of China and Japan are said to contain many allusions to the use of natural gas for lighting and heating. Petroleum was known as *burning water* in Japan in the 7th century. In his book *Dream Pool Essays* written in 1088, the polymathic scientist and statesman Shen Kuo of the Song Dynasty coined the word 石油 (*Shíyóu*, literally "rock oil") for petroleum, which remains the term used in contemporary Chinese.

In southwest Asia the first streets of 8th century Baghdad were paved with tar, derived from natural seep fields in the region. In the 9th century, oil fields were exploited in the area around modern Baku, Azerbaijan. These fields were described by the Arab geographer Abu al-Hasan ʿAlī al-Masʿūdī in the 10th century, and by Marco Polo in the 13th century, who described the output of those wells as hundreds of shiploads. Distillation of petroleum was described by the Persian alchemist, Muhammad ibn Zakarīya Rāzi (Rhazes). There was production of chemicals such as kerosene in the alembic (*al-ambiq*), which was mainly used for kerosene lamps. Arab and Persian chemists also distilled crude oil in order to produce flammable products for military purposes. Through Islamic Spain, distillation became available in Western Europe by the 12th century. It has also been present in Romania since the 13th century, being recorded as păcură.

Eighteenth Century Europe

In Europe, petroleum seeps were extensively mined near the Alsace city of Pechelbronn, where the vapor separation process was in use in 1742. In Switzerland c. 1710, the Russian-born Swiss physician and Greek teacher Eyrini d'Eyrinis discovered asphaltum at Val-de-Travers, (Neuchâtel). He established a bitumen mine *de la Presta* there in 1719 that operated until 1986. Oil sands here were mined from 1745 under the direction of Louis Pierre Ancillon de la Sablonnière, by special appointment of Louis XV. The Pechelbronn oil field was active until 1970, and was the birthplace of companies like Antar and Schlumberger. In 1745 under the Empress Elisabeth of Russia the first oil well and refinery were built in Ukhta by Fiodor Priadunov. Through the process of distillation of the "rock oil" (petroleum) he received a kerosene-like substance, which was used in oil lamps by Russian churches and monasteries (though households still relied on candles).

Colonial Americas

The earliest mention of petroleum seeps in the Americas occurs in Sir Walter Raleigh's account of the Pitch Lake on Trinidad in 1595. Thirty-seven years later, the account of a visit of a Franciscan, Joseph de la Roche d'Allion, to the oil springs of New York was published in Sagard's *Histoire du Canada*. In North America, the early European fur traders found Canadian First Nations using bitumen from the vast Athabasca oil sands to waterproof their birch bark canoes. A Swedish scientist, Peter Kalm, in his 1753 work *Travels into North America*, showed on a map the oil springs of Pennsylvania.

In 1769 the Portolà expedition, a group of Spanish explorers led by Gaspar de Portolà, made the first written record of the tar pits in California. Father Juan Crespí wrote, "While crossing the basin the scouts reported having seen some geysers of tar issuing from the ground like springs; it boils up molten, and the water runs to one side and the tar to the other. The scouts reported that they had come across many of these springs and had seen large swamps of them, enough, they said, to caulk many vessels. We were not so lucky ourselves as to see these tar geysers, much though we wished it; as it was some distance out of the way we were to take, the Governor [Portola] did not want us to go past them. We christened them *Los Volcanes de Brea* [the Tar Volcanoes]."

Modern Extraction and Industry

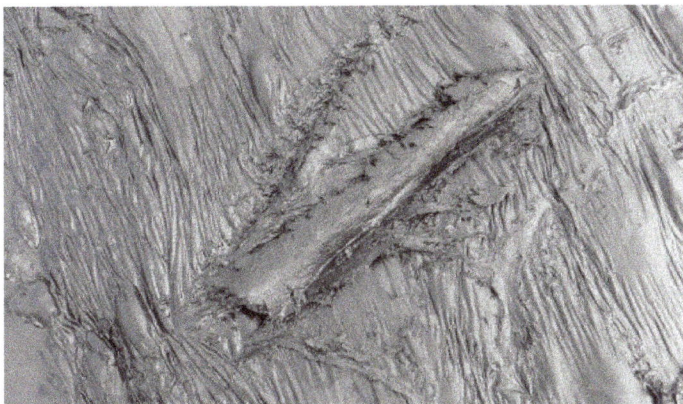

Natural asphalt with embedded wood, Pitch Lake, Trinidad and Tobago. Asphalt has been produced from Pitch Lake since 1851.

During the nineteenth and the beginning of the twentieth century, oil seepages in Europe were exploited everywhere with the digging, and later drilling, of wells near to their occurrences and the discovery of numerous small oil fields such as in Italy.

The modern history of petroleum exploitation, in relation to extraction from seeps, began in the 19th century with the refining of kerosene from crude oil as early as 1823, and the process of refining kerosene from coal by Nova Scotian Abraham Pineo Gesner in 1846. It was only after Ignacy Łukasiewicz had improved Gesner's method to develop a means of refining kerosene from the more readily available "rock oil" ("petr-oleum") seeps in 1852 that the first rock oil mine was built near Krosno in central European Galicia (Poland/Ukraine) in 1853. In 1854, Benjamin Silliman, a science professor at Yale University, was the first American to fractionate petroleum by distillation. These discoveries rapidly spread around the world.

The world's first commercial oil well was drilled in Poland in 1853, and the second in nearby Romania in 1857. At around the same time the world's first, but small, oil refineries were opened at Jasło in Poland, with a larger one being opened at Ploieşti in Romania shortly after. Romania is the first country in the world to have its crude oil output officially recorded in international statistics, namely 275 tonnes. By the end of the 19th century the Russian Empire, particularly in Azerbaijan, had taken the lead in production.

The first oil "well" in North America was in Oil Springs, Ontario, Canada in 1858, dug by James Miller Williams. The US petroleum industry began with Edwin Drake's drilling of a 69-foot (21 m) oil well in 1859 on Oil Creek near Titusville, Pennsylvania, both named for their petroleum seeps.

Other sources of oil initially associated with petroleum seeps were discovered in Peru's Zorritos District in 1863, in the Dutch East Indies on Sumatra in 1885, in Persia at Masjed Soleiman in 1908, as well as in Venezuela, Mexico, and the province of Alberta, Canada.

By 1910, these too were being developed at an industrial level. Initially these petroleum sources and products were for use in fueling lamps, but with the development of the internal combustion engine, their supply could not meet the increased demand; many of these early traditional sources and "local finds" were soon outpaced by technology and demand.

Petroleum Seep Formation

Tar seep at Rozel Point, on the mud (salt) flats of the Great Salt Lake.

Re-worked tar on beach at Rozel Point, Utah. The black rocks are basalt: the tar is brown and looks like cow manure.

A petroleum seep occurs as a result of the seal above the reservoir being breached, causing tertiary migration of hydrocarbons towards the surface under the influence of the associated buoyancy force. The seal is breached due to the effects of overpressure adding to the buoyancy force, overcoming the capillary resistance that initially kept the hydrocarbons sealed.

Causes of Overpressure

The most common cause of overpressure is the rapid loading of fine-grained sediments preventing water from escaping fast enough to equalise the pressure of the overburden. If burial stops or slows, then excess pressure can equalize at a rate that is dependent on the permeability of the overlying and adjacent rocks. A secondary cause of overpressure is fluid expansion, due to changes in the volume of solid and/or fluid phases. Some examples include: aquathermal pressuring (thermal expansion), clay dehydration reactions (such as anhydrite) and mineral transformation (such as kerogen to oil/gas and excess kerogen).

Types of Seeps

There are two types of seep that can occur, depending on the degree of overpressure. Capillary failure can occur in moderate overpressure conditions, resulting in widespread but low intensity seepage until the overpressure equalizes and resealing occurs. In some cases, the moderate overpressure cannot be equalized because the pores in the rock are small so the displacement pressure, the pressure required to break the seal, is very high. If the overpressure continues to increase to the point that it overcomes the rock's minimum stress and its tensile strength before overcoming the displacement pressure, then the rock will fracture, causing local and high intensity seepage until the pressure equalizes and the fractures close.

California Seeps

California has several hundred naturally occurring seeps, found in 28 counties across the state. Much of the petroleum discovered in California during the 19th century was from observations of seeps. The world's largest natural oil seepage is Coal Oil Point in the Santa Barbara Channel, California. Three of the better known tar seep locations in California are McKittrick Tar Pits, Carpinteria Tar Pits and the La Brea Tar Pits.

Diatomite outcrop containing oil that seeps out in hot weather, near McKittrick, in Kern County California.

Oil stained outcrop near Kern River oilfield, in Kern County California.

At Kern River Oil Field, there are no currently active seeps. However, oil-stained formations in the outcrops remain from previously active seeps. Petroleum seeps may be a significant source of pollution.

Seeps known as the McKittrick Tar Pits occur in the McKittrick Oil Field in western Kern County. Some of the seeps occur in watersheds that drain toward the San Joaquin Valley floor. These seeps were originally mined for asphalt by Native Americans, and in the 1870s larger scale mining was undertaken by means of both open pits and shafts. In 1893, Southern Pacific Railroad constructed a line to Asphalto, two miles from present day McKittrick. Fuel oil for the railroad was highly desired, especially since there are very few coal-bearing formations in California. The field is produced now by conventional oil wells, as well as by steam fracturing.

The oil seeps at McKittrick are located in diatomite formation that has been thrust faulted over the younger sandstone formations. Similarly, in the Upper Ojai Valley in Ventura County, tar seeps are aligned with east-west faulting. In the same area, Sulphur Mountain is named for the hydrogen sulfide-laden springs. The oil fields in the Sulphur Mountain area date from the 1870s. Production was from tunnels dug into the face of a cliff, and produced by gravity drainage.

The petroleum fly (*Helaeomyia petrolei*) is a species of fly that was first described from the La Brea Tar Pits and is found at other California seeps as well. It is highly unusual among insects for its tolerance of crude oil; larvae of this fly live within petroleum seeps where they feed on insects and other arthropods that die after becoming trapped in the oil.

Offshore Seeps

In the Gulf of Mexico, there are more than 600 natural oil seeps that leak between one and five million barrels of oil per year, equivalent to roughly 80,000 to 200,000 tonnes. When a petroleum seep forms underwater it may form a peculiar type of volcano known as an asphalt volcano.

The California Division of Oil, Gas and Geothermal Resources published a map of offshore oil seeps from Point Aguello (north of Santa Barbara) to Mexico. In addition, they published a brochure describing the seeps. The brochure also discusses the underground blowout at Platform A which caused the 1969 Santa Barbara oil spill. It also describes accounts from divers, who describe seepage changes after the 1971 San Fernando earthquake.

In Utah, there are natural oil seeps at Rozel Point on the Great Salt Lake. The oil seeps at Rozel Point can be seen when the lake level drops below an elevation of approximately 4,198 feet (1,280 m); if the lake level is higher, the seeps are underwater. The seeps can be found by going to the Golden Spike historical site, and from there, following signs for the Spiral Jetty. Both fresh tar seeps and re-worked tar (tar caught by the waves and thrown up on the rocks) are visible at the site.

The petroleum seeping at Rozel Point is high in sulfur, but has no hydrogen sulfide. This may be related to deposition in a hypersaline lacustrine environment.

Abiogenic Petroleum Origin

Abiogenic petroleum origin is a term used to describe a number of different hypotheses which propose that petroleum and natural gas are formed by inorganic means rather than by the decomposition of organisms. The two principal abiogenic petroleum hypotheses, the deep gas hypothesis of Thomas Gold and the deep abiotic petroleum hypothesis, have been scientifically discredited and are obsolete. Scientific opinion on the origin of oil and gas is that all natural oil and gas deposits on Earth are fossil fuels, and are therefore not abiogenic in origin. Abiogenesis of small quantities of oil and gas remains a minor area of ongoing research.

Some abiogenic hypotheses have proposed that oil and gas did not originate from fossil deposits, but have instead originated from deep carbon deposits, present since the formation of the Earth. Additionally, it has been suggested that hydrocarbons may have arrived on Earth from solid bodies such as comets and asteroids from the late formation of the Solar System, carrying hydrocarbons with them.

Some abiogenic hypotheses gained limited popularity among geologists over the past several centuries. Scientists in the former Soviet Union widely held that significant petroleum deposits could be attributed to abiogenic origin, though this view fell out of favor toward the end of the 20th century because they did not make useful predictions for the discovery of oil deposits. Previous to 2016, it was generally accepted that abiogenic formation of petroleum has insufficient scientific support and that oil and gas fuels on Earth are formed almost exclusively from organic material.

The abiogenic hypothesis regained support in 2009 when researchers at the Royal Institute of

Technology (KTH) in Stockholm reported they believed they had proven that fossils from animals and plants are not necessary for crude oil and natural gas to be generated.

History

An abiogenic hypothesis was first proposed by Georgius Agricola in the 16th century and various additional abiogenic hypotheses were proposed in the 19th century, most notably by Prussian geographer Alexander von Humboldt, the Russian chemist Dmitri Mendeleev and the French chemist Marcellin Berthelot. Abiogenic hypotheses were revived in the last half of the 20th century by Soviet scientists who had little influence outside the Soviet Union because most of their research was published in Russian. The hypothesis was re-defined and made popular in the West by Thomas Gold who published all his research in English.

Abraham Gottlob Werner and the proponents of neptunism in the 18th century regarded basaltic sills as solidified oils or bitumen. While these notions proved unfounded, the basic idea of an association between petroleum and magmatism persisted. Alexander von Humboldt proposed an inorganic abiogenic hypothesis for petroleum formation after he observed petroleum springs in the Bay of Cumaux (Cumaná) on the northeast coast of Venezuela. He is quoted as saying in 1804, "the petroleum is the product of a distillation from great depth and issues from the primitive rocks beneath which the forces of all volcanic action lie". Other prominent proponents of what would become the abiogenic hypothesis included Mendeleev (1877) and Berthelot (1827-1907).

In 1951, the Soviet geologist Nikolai Alexandrovitch Kudryavtsev proposed the modern abiotic hypothesis of petroleum. On the basis of his analysis of the Athabasca Oil Sands in Alberta, Canada, he concluded that no "source rocks" could form the enormous volume of hydrocarbons, and therefore offered abiotic deep petroleum as the most plausible explanation. (Humic coals have since been proposed for the source rocks.) Others who continued Kudryavtsev's work included Petr N. Kropotkin, Vladimir B. Porfir'ev, Emmanuil B. Chekaliuk, Vladilen A. Krayushkin, Georgi E. Boyko, Georgi I. Voitov, Grygori N. Dolenko, Iona V. Greenberg, Nikolai S. Beskrovny, and Victor F. Linetsky.

Astronomer Thomas Gold was a prominent proponent of the abiogenic hypothesis in the West until his death in 2004. More recently, Jack Kenney of Gas Resources Corporation has come to prominence.

State of Current Research

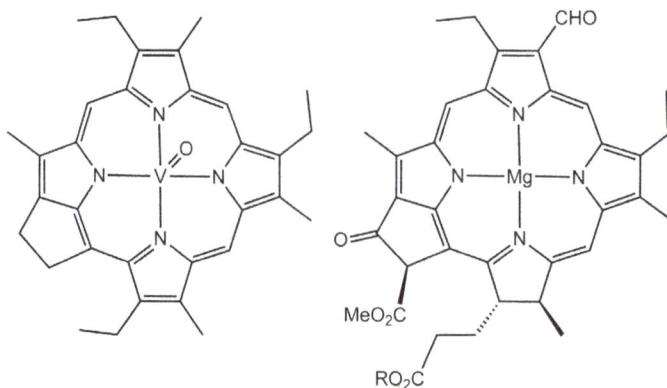

Structure of a biomarker extracted from petroleum and simplified structure of chlorophyll a.

The weight of evidence currently shows that petroleum is derived from ancient biomass. However, it still has to be established conclusively, which means that abiogenic alternative theories of petroleum formation cannot be dismissed.

A 2006 review article by Glasby presented arguments against the abiogenic origin of petroleum on a number of counts.

Foundations of Abiogenic Theories

Within the mantle, carbon may exist as hydrocarbons—chiefly methane—and as elemental carbon, carbon dioxide, and carbonates. The abiotic hypothesis is that the full suite of hydrocarbons found in petroleum can either be generated in the mantle by abiogenic processes, or by biological processing of those abiogenic hydrocarbons, and that the source-hydrocarbons of abiogenic origin can migrate out of the mantle into the crust until they escape to the surface or are trapped by impermeable strata, forming petroleum reservoirs.

Abiogenic hypotheses generally reject the supposition that certain molecules found within petroleum, known as biomarkers, are indicative of the biological origin of petroleum. They contend that these molecules mostly come from microbes feeding on petroleum in its upward migration through the crust, that some of them are found in meteorites, which have presumably never contacted living material, and that some can be generated abiogenically by plausible reactions in petroleum.

Some of the evidence used to support abiogenic theories includes:

Proponents	Item
Gold	The presence of methane on other planets, meteors, moons and comets
Gold, Kenney	Proposed mechanisms of abiotically chemically synthesizing hydrocarbons within the mantle
Kudryavtsev, Gold	Hydrocarbon-rich areas tend to be hydrocarbon-rich at many different levels
Kudryavtsev, Gold	Petroleum and methane deposits are found in large patterns related to deep-seated large-scale structural features of the crust rather than to the patchwork of sedimentary deposits
Gold	Interpretations of the chemical and isotopic composition of natural petroleum
Kudryavtsev, Gold	The presence of oil and methane within non-sedimentary rocks upon the Earth
Gold	The existence of methane hydrate deposits
Gold	Perceived ambiguity in some assumptions and key evidence used in the conventional understanding of petroleum origin.
Gold	Bituminous coal creation is based upon deep hydrocarbon seeps
Gold	Surface carbon budget and oxygen levels stable over geologic time scales
Kudryavtsev, Gold	The biogenic explanation does not explain some hydrocarbon deposit characteristics
Szatmari	The distribution of metals in crude oils fits better with upper serpentinized mantle, primitive mantle and chondrite patterns than oceanic and continental crust, and show no correlation with sea water
Gold	The association of hydrocarbons with helium, a noble gas

Recent Investigation of Abiogenic Theories

As of 2009, little research is directed on establishing abiogenic petroleum or methane, although the Carnegie Institution for Science has reported that ethane and heavier hydrocarbons can be synthesized under conditions of the upper mantle. Research mostly related to astrobiology and the deep microbial biosphere and serpentinite reactions, however, continue to provide insight into the contribution of abiogenic hydrocarbons into petroleum accumulations.

- rock porosity and migration pathways for abiogenic petroleum

- ocean floor hydrothermal vents as in the Lost City hydrothermal field;

- Mud volcanoes and the volatile contents of deep pelagic oozes and deep formation brines

- mantle peridotite serpentinization reactions and other natural Fischer-Tropsch analogs

- Primordial hydrocarbons in meteorites, comets, asteroids and the solid bodies of the Solar System

    o Primordial or ancient sources of hydrocarbons or carbon in Earth

        ☐ Primordial hydrocarbons formed from hydrolysis of metal carbides of the iron peak of cosmic elemental abundance (chromium, iron, nickel, vanadium, manganese, cobalt)

- isotopic studies of groundwater reservoirs, sedimentary cements, formation gases and the composition of the noble gases and nitrogen in many oil fields

- the geochemistry of petroleum and the presence of trace metals related to Earth's mantle (nickel, vanadium, cadmium, arsenic, lead, zinc, mercury and others)

Similarly, research into the deep microbial hypothesis of hydrocarbon generation is advancing as part of the attempt to investigate the concept of panspermia and astrobiology, specifically using deep microbial life as an analog for life on Mars. Research applicable to deep microbial petroleum theories includes

- Research into how to sample deep reservoirs and rocks without contamination

- Sampling deep rocks and measuring chemistry and biological activity

- Possible energy sources and metabolic pathways which may be used in a deep biosphere

- Investigations into the reworking primordial hydrocarbons by bacteria and their effects on carbon isotope fractionation

Proposed Mechanisms of Abiogenic Petroleum

Primordial Deposits

Thomas Gold's work was focused on hydrocarbon deposits of primordial origin. Meteorites are believed to represent the major composition of material from which the Earth was formed. Some meteorites, such as carbonaceous chondrites, contain carbonaceous material. If a large amount of this material is still within the Earth, it could have been leaking upward for billions of years. The thermodynamic conditions within the mantle would allow many hydrocarbon molecules to be at equilibrium under high pressure and high temperature. Although molecules in these conditions may disassociate, resulting fragments would be reformed due to the pressure. An average equilibrium of various molecules would exist depending upon conditions and the carbon-hydrogen ratio of the material.

Creation within the Mantle

Russian researchers concluded that hydrocarbon mixes would be created within the mantle. Experiments under high temperatures and pressures produced many hydrocarbons—including n-alkanes through $C_{10}H_{22}$—from iron oxide, calcium carbonate, and water. Because such materials are in the mantle and in subducted crust, there is no requirement that all hydrocarbons be produced from primordial deposits.

Hydrogen Generation

Hydrogen gas and water have been found more than 6,000 metres (20,000 ft) deep in the upper crust in the Siljan Ring boreholes and the Kola Superdeep Borehole. Data from the western United States suggests that aquifers from near the surface may extend to depths of 10,000 metres (33,000 ft) to 20,000 metres (66,000 ft). Hydrogen gas can be created by water reacting with silicates, quartz, and feldspar at temperatures in the range of 25 °C (77 °F) to 270 °C (518 °F). These minerals are common in crustal rocks such as granite. Hydrogen may react with dissolved carbon compounds in water to form methane and higher carbon compounds.

One reaction not involving silicates which can create hydrogen is:

Ferrous oxide + water → magnetite + hydrogen

$$3FeO + H_2O \rightarrow Fe_3O_4 + H_2$$

The above reaction operates best at low pressures. At pressures greater than 5 gigapascals (49,000 atm) almost no hydrogen is created.

Thomas Gold reported that hydrocarbons were found in the Siljan Ring borehole and in general increased with depth, although the venture was not a commercial success.

However, several geologists analysed the results and said that no hydrocarbon was found.

Evidence of Abiogenic Mechanisms

- Theoretical calculations by J.F. Kenney using scaled particle theory (a statistical mechanical model) for a simplified perturbed hard-chain predict that methane compressed to 30,000 bars (3.0 GPa) or 40,000 bars (4.0 GPa) kbar at 1,000 °C (1,830 °F) (conditions in the mantle) is relatively unstable in relation to higher hydrocarbons. However, these calculations do not include methane pyrolysis yielding amorphous carbon and hydrogen, which is recognized as the prevalent reaction at high temperatures.

- Experiments in diamond anvil high pressure cells have resulted in partial conversion of methane and inorganic carbonates into light hydrocarbons.,

Biotic (Microbial) Hydrocarbons

The "deep biotic petroleum hypothesis", similar to the abiogenic petroleum origin hypothesis, holds that not all petroleum deposits within the Earth's rocks can be explained purely according to the orthodox view of petroleum geology. Thomas Gold used the term *the deep hot biosphere* to describe the microbes which live underground.

This hypothesis is different from biogenic oil in that the role of deep-dwelling microbes is a biological source for oil which is not of a sedimentary origin and is not sourced from surface carbon. Deep microbial life is only a contaminant of primordial hydrocarbons. Parts of microbes yield molecules as biomarkers.

Deep biotic oil is considered to be formed as a byproduct of the life cycle of deep microbes. Shallow biotic oil is considered to be formed as a byproduct of the life cycles of shallow microbes.

Microbial Biomarkers

Thomas Gold, in a 1999 book, cited the discovery of thermophile bacteria in the Earth's crust as new support for the postulate that these bacteria could explain the existence of certain biomarkers in extracted petroleum. A rebuttal of biogenic origins based on biomarkers has been offered by Kenney, et al. (2001).

Isotopic Evidence

Methane is ubiquitous in crustal fluid and gas. Research continues to attempt to characterise crustal sources of methane as biogenic or abiogenic using carbon isotope fractionation of observed gases (Lollar & Sherwood 2006). There are few clear examples of abiogenic methane-ethane-butane, as the same processes favor enrichment of light isotopes in all chemical reactions, whether organic or inorganic. $\delta^{13}C$ of methane overlaps that of inorganic carbonate and graphite in the crust, which are heavily depleted in ^{12}C, and attain this by isotopic fractionation during metamorphic reactions.

One argument for abiogenic oil cites the high carbon depletion of methane as stemming from the observed carbon isotope depletion with depth in the crust. However, diamonds, which are definitively of mantle origin, are not as depleted as methane, which implies that methane carbon isotope fractionation is not controlled by mantle values.

Commercially extractable concentrations of helium (greater than 0.3%) are present in natural gas from the Panhandle-Hugoton fields in the USA, as well as from some Algerian and Russian gas fields.

Helium trapped within most petroleum occurrences, such as the occurrence in Texas, is of a distinctly crustal character with an Ra ratio of less than 0.0001 that of the atmosphere.

The Chimaera gas seep, near Antalya (SW Turkey), new and thorough molecular and isotopic analyses including methane (~87% v/v; D13C1 from -7.9 to -12.3 ‰; D13D1 from -119 to -124 ‰), light alkanes (C2+C3+C4+C5 = 0.5%; C6+: 0.07%; D13C2 from -24.2 to -26.5 ‰; D13C3 from -25.5 to -27 ‰), hydrogen (7.5 to 11%), carbon dioxide (0.01-0.07%; D13CCO2: -15 ‰), helium (~80 ppmv; R/Ra: 0.41) and nitrogen (2-4.9%; D15N from -2 to -2.8 ‰) converge to indicate that the seep releases a mixture of organic thermogenic gas, related to mature Type III kerogen occurring in Paleozoic and Mesozoic organic rich sedimentary rocks, and abiogenic gas produced by low temperature serpentinization in the Tekirova ophiolitic unit.

Biomarker Chemicals

Certain chemicals found in naturally occurring petroleum contain chemical and structural similarities to compounds found within many living organisms. These include terpenoids, terpenes, pristane, phytane, cholestane, chlorins and porphyrins, which are large, chelating molecules in the same family as heme and chlorophyll. Materials which suggest certain biological processes include tetracyclic diterpane and oleanane.

The presence of these chemicals in crude oil is a result of the inclusion of biological material in the oil; these chemicals are released by kerogen during the production of hydrocarbon oils, as these are chemicals highly resistant to degradation and plausible chemical paths have been studied. Abiotic defenders state that biomarkers get into oil during its way up as it gets in touch with ancient fossils. However a more plausible explanation is that biomarkers are traces of biological molecules from bacteria (archaea) that feed on primordial hydrocarbons and die in that environment. For example, hopanoids are just parts of the bacterial cell wall present in oil as contaminant.

Trace Metals

Nickel (Ni), vanadium (V), lead (Pb), arsenic (As), cadmium (Cd), mercury (Hg) and others metals frequently occur in oils. Some heavy crude oils, such as Venezuelan heavy crude have up to 45% vanadium pentoxide content in their ash, high enough that it is a commercial source for vanadium. Abiotic supporters argue that these metals are common in Earth's mantle, but relatively high contents of nickel, vanadium, lead and arsenic can be usually found in almost all marine sediments.

Analysis of 22 trace elements in oils correlate significantly better with chondrite, serpentinized fertile mantle peridotite, and the primitive mantle than with oceanic or continental crust, and shows no correlation with seawater.

Reduced Carbon

Sir Robert Robinson studied the chemical makeup of natural petroleum oils in great detail, and concluded that they were mostly far too hydrogen-rich to be a likely product of the decay of plant debris, assuming a dual origin for Earth hydrocarbons. However, several processes which generate hydrogen could supply kerogen hydrogenation which is compatible with the conventional explanation.

Olefins, the unsaturated hydrocarbons, would have been expected to predominate by far in any material that was derived in that way. He also wrote: "Petroleum ... [seems to be] a primordial hydrocarbon mixture into which bio-products have been added."

This has however been demonstrated later to be a misunderstanding by Robinson, related to the fact that only short duration experiments were available to him. Olefins are thermally very unstable (that is why natural petroleum normally does not contain such compounds) and in laboratory experiments that last more than a few hours, the olefins are no longer present.

The presence of low-oxygen and hydroxyl-poor hydrocarbons in natural living media is supported by the presence of natural waxes (n=30+), oils (n=20+) and lipids in both plant matter and animal matter, for instance fats in phytoplankton, zooplankton and so on. These oils and waxes, however, occur in quantities too small to significantly affect the overall hydrogen/carbon ratio of biological materials. However, after the discovery of highly aliphatic biopolymers in algae, and that oil generating kerogen essentially represent concentrates of such materials, no theoretical problem exists anymore. Also, the millions of source rock samples that have been analyzed for petroleum yield by the petroleum industry have confirmed the large quantities of petroleum found in sedimentary basins.

Empirical Evidence

Occurrences of abiotic petroleum in commercial amounts in the oil wells in offshore Vietnam are sometimes cited, as well as in the Eugene Island block 330 oil field, and the Dnieper-Donets Basin. However, the origins of all these wells can also be explained with the biotic theory. Modern geologists think that commercially profitable deposits of abiotic petroleum *could* be found, but no current deposit has convincing evidence that it originated from abiotic sources.

The Soviet school saw evidence of their hypothesis in the fact that some oil reservoirs exist in non-sedimentary rocks such as granite, metamorphic or porous volcanic rocks. However, opponents noted that non-sedimentary rocks served as reservoirs for biologically originated oil expelled from nearby sedimentary source rock through common migration or re-migration mechanisms.

The following observations have been commonly used to argue for the abiogenic hypothesis, however each observation of actual petroleum can also be fully explained by biotic origin:

Lost City Hydrothermal Vent Field

The Lost City hydrothermal field was determined to have abiogenic hydrocarbon production. Proskurowski et al. wrote, "Radiocarbon evidence rules out seawater bicarbonate as the carbon source for FTT reactions, suggesting that a mantle-derived inorganic carbon source is leached from the host rocks. Our findings illustrate that the abiotic synthesis of hydrocarbons in nature may occur in the presence of ultramafic rocks, water, and moderate amounts of heat."

Siljan Ring Crater

The Siljan Ring meteorite crater, Sweden, was proposed by Thomas Gold as the most likely place to test the hypothesis because it was one of the few places in the world where the granite basement was cracked sufficiently (by meteorite impact) to allow oil to seep up from the mantle; furthermore it is infilled with a relatively thin veneer of sediment, which was sufficient to trap any abiogenic oil, but was modelled as not having been subjected to the heat and pressure conditions (known as the "oil window") normally required to create biogenic oil. However, some geochemists concluded by geochemical analysis that the oil in the seeps came from the organic-rich Ordovician Tretaspis shale, where it was heated by the meteorite impact.

In 1986–1990 The Gravberg-1 borehole was drilled through the deepest rock in the Siljan Ring in which proponents had hoped to find hydrocarbon reservoirs. It stopped at the depth of 6,800 metres (22,300 ft) due to drilling problems, after private investors spent $40 million. Some eighty barrels of magnetite paste and hydrocarbon-bearing sludge were recovered from the well; Gold maintained that the hydrocarbons were chemically different from, and not derived from, those added to the borehole, but analyses showed that the hydrocarbons were derived from the diesel fuel-based drilling fluid used in the drilling. This well also sampled over 13,000 feet (4,000 m) of methane-bearing inclusions.

In 1991–1992, a second borehole, Stenberg-1, was drilled a few miles away to a depth of 6,500 metres (21,300 ft), finding similar results. Again, no abiotic hydrocarbons were found.

Bacterial Mats

Direct observation of bacterial mats and fracture-fill carbonate and humin of bacterial origin in deep boreholes in Australia are also taken as evidence for the abiogenic origin of petroleum.

Example Proposed Abiogenic Methane Deposits

Panhandle-Hugoton field (Anadarko Basin) in the south-central United States is the most important gas field with commercial helium content. Some abiogenic proponents interpret this as evidence that both the helium and the natural gas came from the mantle.

The Bạch Hổ oil field in Vietnam has been proposed as an example of abiogenic oil because it is 4,000 m of fractured basement granite, at a depth of 5,000 m. However, others argue that it contains biogenic oil which leaked into the basement horst from conventional source rocks within the Cuu Long basin.

A major component of mantle-derived carbon is indicated in commercial gas reservoirs in the Pannonian and Vienna basins of Hungary and Austria.

Natural gas pools interpreted as being mantle-derived are the Shengli Field and Songliao Basin, northeastern China.

The Chimaera gas seep, near Çıralı, Antalya (southwest Turkey), has been continuously active for millennia and it is known to be the source of the first Olympic fire in the Hellenistic period. On the basis of chemical composition and isotopic analysis, the Chimaera gas is said to be about half biogenic and half abiogenic gas, the largest emission of biogenic methane discovered; deep and pressurized gas accumulations necessary to sustain the gas flow for millennia, posited to be from an inorganic source, may be present. Local geology of Chimaera flames, at exact position of flames, reveals contact between serpentinized ophiolite and carbonate rocks. Fischer-Tropsch process can be suitable reaction to form hydrocarbon gases.

Geological Arguments

Incidental Arguments for Abiogenic Oil

Given the known occurrence of methane and the probable catalysis of methane into higher atomic weight hydrocarbon molecules, various abiogenic theories consider the following to be key observations in support of abiogenic hypotheses:

- the serpentinite synthesis, graphite synthesis and spinel catalysation models prove the process is viable

- the likelihood that abiogenic oil seeping up from the mantle is trapped beneath sediments which effectively seal mantle-tapping faults

- outdated mass-balance calculations for supergiant oilfields which argued that the calculated source rock could not have supplied the reservoir with the known accumulation of oil, implying deep recharge.

- the presence of hydrocarbons encapsulated in diamonds

The proponents of abiogenic oil also use several arguments which draw on a variety of natural phenomena in order to support the hypothesis:

- the modeling of some researchers shows the Earth was accreted at relatively low temperature, thereby perhaps preserving primordial carbon deposits within the mantle, to drive abiogenic hydrocarbon production

- the presence of methane within the gases and fluids of mid-ocean ridge spreading centre hydrothermal fields

- the presence of diamond within kimberlites and lamproites which sample the mantle depths proposed as being the source region of mantle methane (by Gold et al.).

Incidental Arguments Against Abiogenic Oil

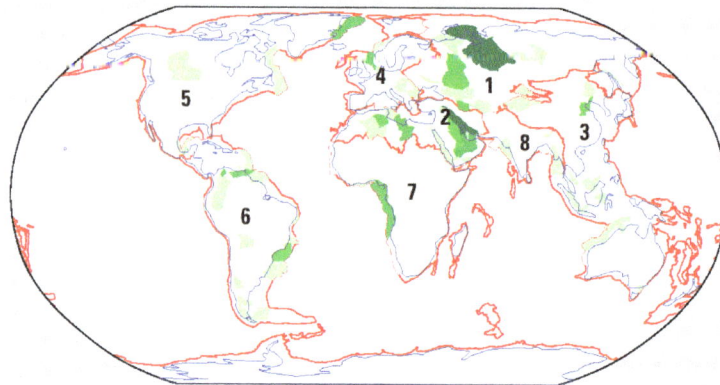

Oil deposits are not directly associated with tectonic structures.

Arguments against chemical reactions, such as the serpentinite mechanism, being the major source of hydrocarbon deposits within the crust include:

- the lack of available pore space within rocks as depth increases

    o this is contradicted by numerous studies which have documented the existence of hydrologic systems operating over a range of scales and at all depths in the continental crust.

- the lack of any hydrocarbon within the crystalline shield areas of the major cratons, especially around key deep seated structures which are predicted to host oil by the abiogenic hypothesis.

- limited evidence that major serpentinite belts underlie continental sedimentary basins which host oil

- lack of conclusive proof that carbon isotope fractionation observed in crustal methane sources is entirely of abiogenic origin (Lollar et al. 2006)

- mass balance problems of supplying enough carbon dioxide to serpentinite within the metamorphic event before the peridotite is fully reacted to serpentinite

- drilling of the Siljan Ring failed to find commercial quantities of oil, thus providing a counter example to Kudryavtsev's Rule and failing to locate the predicted abiogenic oil.

- helium in the Siljan Gravberg-1 well was depleted in ^{3}He and not consistent with a mantle origin

    o The Gravberg-1 well only produced 84 barrels (13.4 m^3) of oil, which later was shown to derive from organic additives, lubricants and mud used in the drilling process.

- the distribution of sedimentary basins is caused by plate tectonics, with sedimentary basins forming on either side of a volcanic arc, which explains the distribution of oil within these sedimentary basins

- Kudryavtsev's Rule has been explained for oil and gas (not coal)—gas deposits which are below oil deposits can be created from that oil or its source rocks. Because natural gas is less dense than oil, as kerogen and hydrocarbons are generating gas the gas fills the top of the available space. Oil is forced down, and can reach the spill point where oil leaks around the edge(s) of the formation and flows upward. If the original formation becomes completely filled with gas then all the oil will have leaked above the original location.

- ubiquitous diamondoids in natural hydrocarbons such as oil, gas and condensates are composed of carbon from biological sources, unlike the carbon found in normal diamonds.

- Gas ruptures during earthquakes are more likely to be sourced from biogenic methane generated in unconsolidated sediment from existing organic matter, released by earthquake liquefaction of the reservoir during tremors

- The presence of methane hydrate is arguably produced by bacterial action upon organic detritus falling from the littoral zone and trapped in the depth due to pressure and temperature

- The likelihood of vast concentrations of methane in the mantle is very slim, given mantle xenoliths have negligible methane in their fluid inclusions; conventional plate tectonics explains deep focus quakes better, and the extreme confining pressures invalidate the hypothesis of gas pockets causing quakes

Extraterrestrial Argument

The presence of methane on Saturn's moon Titan and in the atmospheres of Jupiter, Saturn, Uranus and Neptune is cited as evidence of the formation of hydrocarbons without biological intervention, for example by Thomas Gold. (Terrestrial natural gas is composed primarily of methane). Some comets contain massive amounts of an organic compounds, the equivalent of cubic kilometers of such mixed with other material; for instance, corresponding hydrocarbons were detected during a probe flyby through the tail of Comet Halley in 1986.

Global Strategic Petroleum Reserves

Global strategic petroleum reserves (GSPR) refer to crude oil inventories (or stockpiles) held by the government of a particular country, as well as private industry, to safeguard the economy and help maintain national security during an energy crisis.

According to the United States Energy Information Administration, approximately 4.1 billion barrels (650,000,000 m³) of oil are held in strategic reserves, of which 1.4 billion is government-controlled. The remainder is held by private industry. In 2004 the U.S. Strategic Petroleum Reserve had the largest strategic reserve, with much of the remainder held by the other 27 members of the International Energy Agency. Some non-IEA countries have started work on their own strategic petroleum reserves. China has the largest of these new reserves.

Global oil consumption is in the region of 0.1 billion barrels (16,000,000 m³) per day. The 4.1 bbl reserve is equivalent to 41 days of production. If there ever is a dramatic fall in global output, as envisaged by some peak oil analysts, the strategic petroleum reserves might be used to cover the shortfall. Covering a 50% shortfall would deplete the reserves in 82 days, although export leaders the Middle East and Russian exports represent only 22% and 6% respectively of global production.

International Energy Agency reserves

According to a March 2001 agreement, all 28 members of the International Energy Agency must have a strategic petroleum reserve equal to 90 days of the previous year's net oil imports for their respective countries. Only net-exporter members of the IEA are exempt from this requirement. The exempt countries are Canada, Denmark, Norway, and the United Kingdom. However, the UK and Denmark recently drew up plans to create their own strategic reserves in order to meet their legal obligations as European Union member states.

Forward Commercial Storage Agreements

To allow oil-exporting countries increased flexibility in their production quotas, there has been a progressive movement towards forward commercial storage agreements. These agreements allow petroleum to be stored within an oil-importing country. However, the reserves are technically under the control of the oil-exporting country. Such agreements enable oil-importing countries to access these commercial reserves in a timely and cost effective way.

Emergency Oil Sharing Agreements

Several countries have agreements to share their stockpiles with other countries in the event of an emergency.

The Japan, New Zealand And South Korea Agreement

In 2007, Japan announced a plan to share its strategic reserves with other countries in the region. Negotiations are under way between Japan and New Zealand for an oil-sharing deal whereby Japan sells part of its strategic reserves to New Zealand in the event of an emergency. New Zealand would be required to pay the market price for the oil, plus negotiated option fees for the amount of oil previously held for them by Japan.

South Korea and Japan have agreed to share their oil reserves in the event of an emergency.

The United States and Israel Agreement

According to the 1975 Second Sinai withdrawal document signed by the United States and Israel, in an emergency the U.S. is obligated to make oil available for sale to Israel for a period of up to five years.

The France, Germany, and Italy Agreement

France, Germany and Italy have an oil-sharing agreement in place that allows them to buy oil from each other in the event of an emergency. In 1968, the six members of the European Economic Community – Belgium, France, Germany, Italy, Luxembourg and the Netherlands – agreed to maintain a minimum level of crude oil stocks and oil products corresponding to 65 days' worth of domestic consumption. In 1972, this obligation was raised to 90 days.

Africa

Kenya

Kenya is setting up a Strategic Fuel Reserve, similar to that of cereals. The stocks would be procured by the National Oil Corporation of Kenya and stored by the Kenya Pipeline Company Limited.

Malawi

Malawi is considering creating a 22-day reserve of fuel, which is an expansion from the current five-day reserve. The government is planning to build storage facilities in the provinces of Chipoka and Mchinji as well as Kamuzu International Airport.

South Africa

South Africa has an SPR managed by PetroSA. The main facility is the Saldanha Bay oil storage facility, which is a major transit point for oil shipping. Saldanha Bay's six in-ground concrete storage tanks give the facility a storage capacity of 45,000,000 barrels (7,200,000 m³).

Asia

China

In 2007, China announced the expansion of its crude reserves into a two-part system. China's reserves would consist of a government-controlled strategic reserve complemented by mandated commercial reserves. The government-controlled reserves are being completed in three stages. Phase one consisted of a 101,900,000 barrels (16,200,000 m³) reserve, mostly completed by the end of 2008. The second phase of the government-controlled reserves with an additional

170,000,000 barrels (27,000,000 m³) was to be completed by 2011. Recently, Zhang Guobao, head of the National Energy Administration, stated that there will be a third phase that will expand reserves by 204,000,000 barrels (32,400,000 m³) with the goal of increasing China's SPR to 90 days of supply by 2020.

The planned state reserves of 475,900,000 barrels (75,660,000 m³) together with the planned enterprise reserves of 209,440,000 barrels (33,298,000 m³) will provide around 90 days of consumption or a total of 684,340,000 barrels (108,801,000 m³).

India

In 2003, India started development on a strategic crude oil reserve sized at 37,400,000 barrels (5,950,000 m³), enough to provide two weeks of consumption. Petroleum stocks have been transferred from the Indian Oil Corporation (IndianOil) to the Oil Industry Development Board (OIDB). The OIDB then created the Indian Strategic Petroleum Reserves Ltd (ISPRL) to serve as the controlling government agency for the strategic reserve.

The facilities are located at:

- Mangalore, State of Karnataka. Capacity of 10.995 million barrels (1,748,100 m³).

- Padur village, Udupi in the state of Karnataka. Capacity of 18.7 million barrels (2,970,000 m³).

- Visakhapatnam, State of Andhra Pradesh. Capacity of 1.33 million tonnes.

On 21 December 2011, a senior oil ministry official announced that India was planning to augment its crude reserve capacity to 132 million barrels by 2020.

Japan

Japan Shibushi Oil Stockpile Site

As of 2010, Japan has an SPR composed of the following three types of stockpiles:

- State-controlled reserves of petroleum at 11 different locations totaling 324,000,000 barrels (51,500,000 m³). *All numbers, unless otherwise cited, come from p. 177 of this document:*

- o Tomakomai Eastern Oil Reserve Storage Base – 55 storage tanks, total capacity 34 million barrels (5,400,000 m³).

- o Mutsu-Ogawara Storage Base – 53 storage tanks, total capacity 31 million barrels (4,900,000 m³).

- o Kuji Storage Base – three storage tanks, total capacity 10.5 million barrels (1,670,000 m³).

- o Akita Storage Base – 15 storage tanks, total capacity 23.4 million barrels (3,720,000 m³).

- o Fukui Storage Base – 27 storage tanks, total capacity 17.9 million barrels (2,850,000 m³).

- o Kikuma Underground Petroleum Storage Facility – eight storage tanks, total capacity 8.9 million barrels (1,410,000 m³).

- o Shirashima Storage Facility – eight tankers (4,400,000 barrels (700,000 m³) each), total capacity 35.2 million barrels (5,600,000 m³).

- o Kamigotou Storage Base – seven storage tanks, total capacity 21.45 million barrels (3,410,000 m³).

- o Kushikino Storage Base – three storage tanks, total capacity 10.5 million barrels (1,670,000 m³).

- o Shibushi Storage Base – 40 storage tanks, total capacity 27.6 million barrels (4,390,000 m³).

- o Kagoshima – 4.0 million barrels (640,000 m³). A forward commercial storage facility with Abu Dhabi.

- Private reserves of petroleum held "in accordance with the Petroleum Stockpiling Law" of 129,000,000 barrels (20,500,000 m³).

- Private reserves of petroleum products for another 130,000,000 barrels (21,000,000 m³).

The state-controlled reserves and the privately held stockpiles total about 583,000,000 barrels (92,700,000 m³). The Japanese SPR is run by the Japan Oil, Gas and Metals National Corporation.

South Korea

In South Korea, refineries, specified distributors, and importers, are obliged to hold from 40 days to 60 days of their daily import, sale, or refined production, based on the previous 12 months. At the end of 2010, South Korea possessed a total storage capacity of 286 million barrels (45.5 million cubic meters), composed of 146 mb of South Korea National Oil Corporation's facilities used for government stocks and international joint oil stockpiling, and 140 mb used for industry operation and mandatory industry stocks. South Korea's oil stocks in terms of days of net imports have consistently been above 160 days since January 2009, hitting the country's historical record of 240 days (124 days of government stocks and 117 days of industry stocks) in March 2014.

Others

The Philippines had plans for a National Petroleum Strategic Reserve by 2010 with an approximate size of 30,000,000 barrels (4,800,000 m³).

Singapore does not have any oil reserves.

Taiwan has an SPR with a 1999-reported size of 13,000,000 barrels (2,100,000 m³). Taiwan's refiners (Kaohsiung 270,000 bbl/d (43,000 m³/d); Ta-Lin 300,000 bbl/d (48,000 m³/d); Tao-Yuan 200,000 bbl/d (32,000 m³/d); Mailiao 150,000 bbl/d) are also required to store at least 30 days of petroleum stocks. As of 2005, these mandated commercial reserves total 27,600,000 barrels (4,390,000 m³) of strategic petroleum stocks.

Thailand increased the size of its SPR from 60 days to 70 days of consumption in 2006.

Pakistan has announced plans for a 20-day emergency reserve.

Europe

European Union

In the European Union, according to Council Directive 68/414/EEC of 20 December 1968, all 28 member states are required to have a strategic petroleum reserve within the territory of the E.U. equal to at least 90 days of average domestic consumption.

The Czech Republic has a four-tank SPR facility in Nelahozeves run by the company CR Mero. The Czech SPR is equal to 100 days of consumption or 20,300,000 barrels (3,230,000 m³).

Denmark has a reserve equal to 81 days of consumption (about 1.4 million tonnes). Not counting reserves held by the military defence.

Finland has an SPR with an approximate size of 62,400,000 barrels (9,920,000 m³).

France has an SPR with an approximate size of 65,000,000 barrels (10,300,000 m³). As of 2000, jet fuel stocks for at least 55 days of consumption were required, with half of those stocks controlled by the *Société Anonyme de Gestion des Stocks de Sécurité* (SAGESS) and the other half controlled by producers.

Germany created the Federal Oil Reserve in 1970, located in the Etzel salt caverns near Wilhelmshaven in northern Germany, with an initial size of 70 million barrels (11,000,000 m³). The current German Federal Oil Reserve and the Erdölbevorratungsverband (EBV) (the German stockholding company) mandates that refiners must keep 90 days of stock on hand, giving Germany an approximate reserve size of 250,000,000 barrels (40,000,000 m³) as of 1997. The German SPR is the largest in Europe.

Hungary has an SPR equal to approximately 90 days of consumption or 11,880,000 barrels (1,889,000 m³).

Ireland has approximately 31 days of oil stocks in Ireland and another nine days of oil stocks held in other EU members states. Additionally, it has stock tickets (contracts with a third party whereby the government has the option of purchasing oil in the event of an emergency) and stocks held by large industry or large consumers. In total, Ireland has approximately 100 days' worth of oil at its disposal.

The Netherlands maintains a stockpile equal to 90 days of net oil imports. In 2013, this was about four million tonnes of oil.

Poland has an SPR equal to approximately 70 days of consumption. Another facility holding 20 days of consumption was completed in 2008. Poland also requires oil companies to maintain reserves sufficient to provide 73 days of consumption.

Portugal has an SPR with an approximate size of 22,440,000 barrels (3,568,000 m³).

Slovakia has an SPR with an approximate size of 748,000 barrels (118,900 m³).

Spain has an SPR with an approximate size of 120,000,000 barrels (19,000,000 m³).

Sweden has an SPR with an approximate size of 13,290,000 barrels (2,113,000 m³).

The United Kingdom recently drew up plans to create its own strategic fuel reserves.

Russia

As of 2011, Russia is accumulating strategic reserves of refined oil products to be held by Rosneftegaz, a state-owned company. The reserves will be held at commercial refineries, Transneft facilities and state reserve facilities. The current planned size is 14,665,982 barrels (2,331,704.8 m³).

Switzerland

Switzerland has SPRs consisting of gas, diesel, jet fuel and heating oil for 4.5 months of consumption. The reserves were created in the 1940s.

Middle East

Iran

In April 2006, the Fars News Agency reported that Iran was planning to create an SPR. The National Iranian Oil Company (NIOC) began construction of 15 crude oil storage tanks with a capacity of 10,000,000 barrels (1,600,000 m³). In August 2008, Iran announced plans to expand the SPR with a new facility on Kharg Island with four tanks holding 1,000,000 barrels (160,000 m³) each. Iran's SPR facilities are:

- Ahwaz – four storage tanks, total capacity 2 million barrels (320,000 m³).

- Omidiyeh – three storage tanks, total capacity 3 million barrels (480,000 m³).

- Goureh – six storage tanks, total capacity 4 million barrels (640,000 m³).

- Sirri Island – one storage tank, total capacity 500,000 barrels (79,000 m³).

- Bahregansar – one storage tank, total capacity 500,000 barrels (79,000 m³).

- Kharg Island – four storage tanks, total capacity 4 million barrels (640,000 m³). (Planned facility, not operational yet.)

Kuwait

Kuwait has a joint stockpile located in South Korea. The deal gives South Korea first rights to purchase the oil. As of 2006, the size of the stockpile is 2 million barrels (320,000 m³).

Others

As of 1975, Israel is believed to have a strategic oil reserve equal to 270 days of consumption.

Jordan has strategic oil reserves equal to 60 days of consumption or 6,240,000 barrels (992,000 m³).

North America

United States

The United States has the world's largest reported Strategic Petroleum Reserve with a total capacity of 727 million barrels. If completely filled, the U.S. SPR could theorctically replace about 60 days of oil imports. The U.S. is estimated to import approximately 12,000,000 barrels per day (1,900,000 m³/d) of crude oil. According to the U.S. Department of Energy, the facilities' maximum flow rate is limited to approximately 4,400,000 barrels per day (700,000 m³/d) when filled to maximum capacity, declining as the reserve is emptied. The reserves are kept in salt caverns located at:

- Bryan Mound – near Freeport, Texas. Capacity of 226,000,000 barrels (35,900,000 m³).

- Big Hill – near Winnie, Texas. Capacity of 160,000,000 barrels (25,000,000 m³).

- West Hackberry – near Lake Charles, Louisiana. Capacity of 219,000,000 barrels (34,800,000 m³).

- Bayou Choctaw – near Baton Rouge, Louisiana. Capacity of 72,000,000 barrels (11,400,000 m³).

- Richton, Mississippi. A planned facility.

The U.S. also has the 2-million-barrels (320,000 m³) Northeast Home Heating Oil Reserve to supply northeast home owners with heating oil if there is a shortage.

California

The state of California is considering the creation of Strategic Fuels Reserve.

Oceania

As of 2008, New Zealand has a strategic reserve with a size of 170,000 tons or 1,200,000 barrels (190,000 m³). Much of this reserve is based upon ticketed option contracts with Australia, Japan, the United Kingdom and the Netherlands, which allow for guaranteed purchases of petroleum in the event of an emergency.

As of 2008, Australia holds three weeks of petroleum, instead of the allotted 90 days that was agreed upon, according to the study 'Liquid Fuel Security' authored by Air Vice-Marshal John Blackburn, AO (retired).

Hydrocarbon Indicator

A hydrocarbon indicator (HCI) or direct hydrocarbon indicator (DHI), is an anomalous seismic attribute value or pattern that could be explained by the presence of hydrocarbons in a oil or gas reservoir.

DHIs are particularly useful in hydrocarbon exploration for reducing the geological risk of exploration wells. Broadly, geophysicists recognize several types of DHI:

- Bright spots: localized amplitudes of greater magnitude than background amplitude values. Equipment prior to the 1970s had the bright spots obscured due to the automatic gain control.

- Flat spots: nearly horizontal reflectors that cross existing stratigraphy, possibly indicating a hydrocarbon fluid level within an oil or gas reservoir.

- Dim spots: low amplitude anomalies.

- Polarity reversals can occur where the capping rock has a slightly lower seismic velocity than the reservoir and the reflection has its sign reversed.

Some geoscientists regard amplitude versus offset anomalies as a type of direct hydrocarbon indicator. For example, the amplitude of a reflection might increase with the angle of incidence, a possible indicator of natural gas.

Types of Hydrocarbon Indicator

Bright Spot

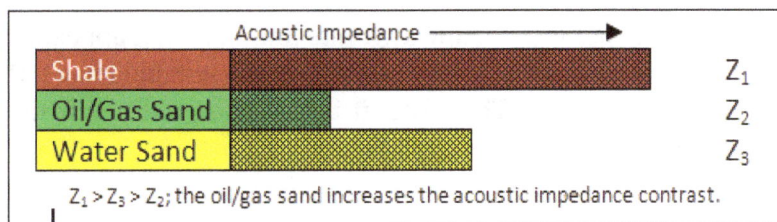

The diagram above shows the acoustic impedance relationship that results in a bright spot.

In reflection seismology, a bright spot is a local high amplitude seismic attribute anomaly that can indicate the presence of hydrocarbons and is therefore known as a direct hydrocarbon indicator. It is used by geophysicists in hydrocarbon exploration.

History

Bright spots were not commonly identified until the early 1970s because of the extensive and industry-wide use of automatic gain control, which obscured the amplitude effects of hydrocarbon accumulations.

Theory

A bright spot primarily results from the increase in acoustic impedance contrast when a hydrocarbon (with a lower acoustic impedance) replaces the brine-saturated zone (with a higher acoustic impedance) that underlies a shale (with a higher acoustic impedance still), increasing the reflection coefficient. The effect decreases with depth because compaction for sands and shales occurs at different rates and the acoustic impedance relationship stated above will not hold after a certain depth/age. Below this depth, there will be a crossover of shale and sand acoustic impedances and a dim spot is more useful to hydrocarbon exploration.

Caution

The relationship between hydrocarbons and direct hydrocarbon indicators, such as bright spots, is not straightforward and not all bright spots are caused by the presence of hydrocarbons and therefore they should not be treated as conclusive evidence of hydrocarbon accumulations.

Flat Spot (Reflection Seismology)

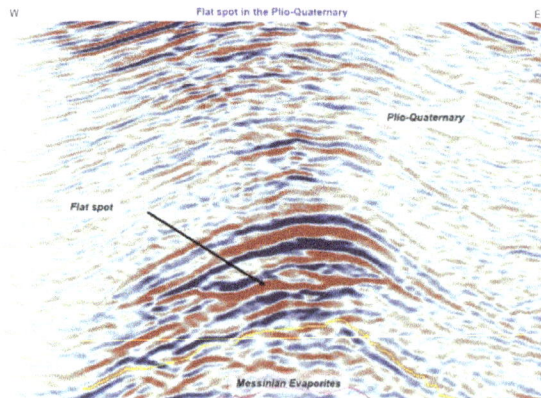

Flat spot in 2D seismic line.

In reflection seismology, a flat spot is a seismic attribute anomaly that appears as a horizontal reflector cutting across the stratigraphy elsewhere present on the seismic image. Its appearance can indicate the presence of hydrocarbons. Therefore, it is known as a direct hydrocarbon indicator and is used by geophysicists in hydrocarbon exploration.

Theory

A flat spot can result from the increase in acoustic impedance when a gas-filled porous rock (with a lower acoustic impedance) overlies a liquid-filled porous rock (with a higher acoustic impedance). It may stand out on a seismic image because it is flat and will contrast with surrounding dipping reflections.

Caution

There are a number of other possible reasons for there being a flat spot on a seismic image. It could be representative of a mineralogical change in the subsurface or an unresolved shallower multiple. Additionally, the interpretation of a flat spot should be attempted after depth conversion to confirm that the anomaly is actually flat.

Dim Spot

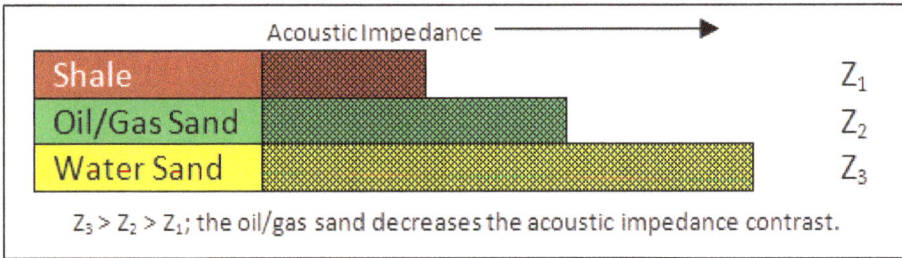

The diagram above shows the acoustic impedance relationship that results in a dim spot.

In reflection seismology, a dim spot is a local low amplitude seismic attribute anomaly that can indicate the presence of hydrocarbons and is therefore known as a direct hydrocarbon indicator. It primarily results from the decrease in acoustic impedance contrast when a hydrocarbon (with a low acoustic impedance) replaces the brine-saturated zone (with a high acoustic impedance) that underlies a shale (with the lowest acoustic impedance of the three), decreasing the reflection coefficient.

Occurrence

For a dim spot to occur, the shale has to have a lower acoustic impedance than both the water sand and the oil/gas sand, which is the opposite situation required for a bright spot to occur. This is possible because compaction causes the acoustic impedances of sands and shales to increase with age and depth but it does not happen uniformly – younger shales have a higher acoustic impedance than younger sands, but this reverses at depth, with older shales having a lower acoustic impedance than older sands.

Similarly to bright spots, not all dim spots are caused by the presence of hydrocarbons and therefore they should not be treated as conclusive evidence hydrocarbon accumulations.

Polarity Reversal (Seismology)

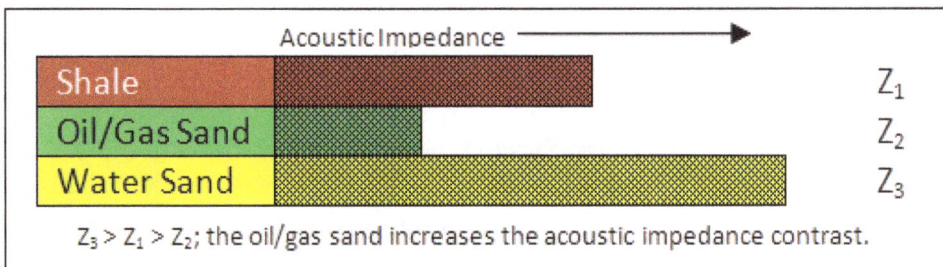

Diagram showing the acoustic relationship that results in a seismic polarity reversal.

In reflection seismology, a polarity reversal or phase change is a local amplitude seismic attribute anomaly that can indicate the presence of hydrocarbons and is therefore known as a direct hydrocarbon indicator. It primarily results from the change in polarity of the seismic response when a shale (with a lower acoustic impedance) overlies a brine-saturated zone (with a high acoustic impedance), that becomes invaded with an oil/gas sand (with the lowest acoustic impedance of the three). This changes the acoustic impedance contrast from an increase to a decrease, resulting in the polarity of the seismic response being reversed - as per the normal convention adopted by the SEG.

Occurrence

For a polarity reversal to occur, the shale has to have a lower acoustic impedance than the water sand and both are required to have a higher acoustic impedance than the oil/gas sand. This is the intermediate situation, that occurs during sediment compaction with depth, between the acoustic impedance relationship required for a bright spot and the acoustic impedance relationship required for a dim spot.

References

- Hyne, Norman J. (2001). Nontechnical Guide to Petroleum Geology, Exploration, Drilling and Production. PennWell Corporation. pp. 431–449. ISBN 9780878148233.

- Lyons, William C. (2005). Standard Handbook Of Petroleum & Natural Gas Engineering. Gulf Professional Publishing. pp. 5–6. ISBN 9780750677851.

- Salim Al-Hassani (2008). "1000 Years of Missing Industrial History". In Emilia Calvo Labarta; Mercè Comes Maymo; Roser Puig Aguilar; Mònica Rius Pinies. A shared legacy: Islamic science East and West. Edicions Universitat Barcelona. pp. 57–82 [63]. ISBN 84-475-3285-2.

- Mackenzie, Sir Alexander (1970). Lamb, W. Kaye, ed. The Journals and Letters of Alexander Mackenzie. Cambridge: Hakluyt Society. p. 129. ISBN 0-521-01034-9.

- Kielbasa, John R. (1998). "Rancho La Brea". Historic Adobes of Los Angeles County. Pittsburg: Dorrance Publishing Co. ISBN 0-8059-4172-X. .

- Akiner, Shirin; Aldis, Anne, eds. (2004). The Caspian: Politics, Energy and Security. New York: Routledge. ISBN 978-0-7007-0501-6. , p. 5

- "Fossils From Animals And Plants Are Not Necessary For Crude Oil And Natural Gas, Swedish Researchers Find.". ScienceDaily. Vetenskapsrådet (The Swedish Research Council). 12 September 2009. Retrieved 9 March 2016.

- George E. Totten. "A Timeline of Highlights from the Histories of ASTM Committee D02 and the Petroleum Industry". ASTM International. Retrieved April 3, 2015.

- "Oil & Gas Seeps - The Natural and Human Histories of Oil and Gas Seeps". California Department of Conservation (Conservation.ca.gov). Retrieved January 24, 2015.

- "A Guide to the Insects of the Coal Oil Point Reserve - Flies". Santa Barbara Museum of Natural History (Sbnature.org). Retrieved January 24, 2015.

- David F. Morehouse (1997). "The Intricate Puzzle of Oil and Gas Reserves Growth" (PDF). U.S. Energy Information Administration. Archived from the original (PDF) on August 6, 2010. Retrieved 2014-08-19.

- Faucon, Benoit (18 July 2011). "Venezuela Oil Reserves Surpassed Saudi Arabia In 2010-OPEC". Fox Business. Retrieved 18 July 2011.

Exploration Techniques

Once a prospect is identified and evaluated, an exploration well is drilled to determine the presence or absence of oil or gas. Exploration geophysics, seismic refraction, gravity gradiometry, electrical resistivity tomography, aeromagnetic survey, magnetotellurics, ground-penetrating radar and transient electromagnetics are the various exploration techniques discussed in the chapter.

Exploration Geophysics

Exploration geophysics is an applied branch of geophysics, which uses physical methods (such as seismic, gravitational, magnetic, electrical and electromagnetic) at the surface of the Earth to measure the physical properties of the subsurface, along with the anomalies in those properties. It is most often used to detect or infer the presence and position of economically useful geological deposits, such as ore minerals; fossil fuels and other hydrocarbons; geothermal reservoirs; and groundwater reservoirs.

Exploration geophysics can be used to directly detect the target style of mineralization, via measuring its physical properties directly. For example, one may measure the density contrasts between iron ore and silicate wall rocks, or may measure the electrical conductivity contrast between conductive sulfide minerals and barren silicate minerals.

Geophysical Methods

The main techniques used are:

1. Seismic methods, such as reflection seismology, seismic refraction, and seismic tomography.

2. Geodesy and gravity techniques, including gravity gradiometry.

3. Magnetic techniques, including aeromagnetic surveys.

4. Electrical techniques, including electrical resistivity tomography and induced polarization.

5. Electromagnetic methods, such as magnetotellurics, ground penetrating radar and transient/time-domain electromagnetics.

6. Borehole geophysics, also called well logging.

7. Remote sensing techniques, including hyperspectral imaging.

Many other techniques, or methods of integration of the above techniques, have been developed and are currently used. However these are not as common due to cost-effectiveness, wide applicability and/or uncertainty in the results produced.

Uses

Exploration geophysics is also used to map the subsurface structure of a region, to elucidate the underlying structures, spatial distribution of rock units, and to detect structures such as faults, folds and intrusive rocks. This is an indirect method for assessing the likelihood of ore deposits or hydrocarbon accumulations.

Methods devised for finding mineral or hydrocarbon deposits can also be used in other areas such as monitoring environmental impact, imaging subsurface archaeological sites, ground water investigations, subsurface salinity mapping, civil engineering site investigations and interplanetary imaging.

Mineral Exploration

Magnetometric surveys can be useful in defining magnetic anomalies which represent ore (direct detection), or in some cases gangue minerals associated with ore deposits (indirect or inferential detection).

The most direct method of detection of ore via magnetism involves detecting iron ore mineralisation via mapping magnetic anomalies associated with banded iron formations which usually contain magnetite in some proportion. Skarn mineralisation, which often contains magnetite, can also be detected though the ore minerals themselves would be non-magnetic. Similarly, magnetite, hematite and often pyrrhotite are common minerals associated with hydrothermal alteration, and this alteration can be detected to provide an inference that some mineralising hydrothermal event has affected the rocks.

Gravity surveying can be used to detect dense bodies of rocks within host formations of less dense wall rocks. This can be used to directly detect Mississippi Valley Type ore deposits, IOCG ore deposits, iron ore deposits, skarn deposits and salt diapirs which can form oil and gas traps.

Electromagnetic (EM) surveys can be used to help detect a wide variety of mineral deposits, especially base metal sulphides via detection of conductivity anomalies which can be generated around sulphide bodies in the subsurface. EM surveys are also used in diamond exploration (where the kimberlite pipes tend to have lower resistance than enclosing rocks), graphite exploration, palaeochannel-hosted uranium deposits (which are associated with shallow aquifers, which often respond to EM surveys in conductive overburden). These are indirect inferential methods of detecting mineralisation, as the commodity being sought is not directly conductive, or not sufficiently conductive to be measurable. EM surveys are also used in unexploded ordnance, archaeological, and geotechnical investigations.

Regional EM surveys are conducted via airborne methods, using either fixed-wing aircraft or helicopter-borne EM rigs. Surface EM methods are based mostly on Transient EM methods using surface loops with a surface receiver, or a downhole tool lowered into a borehole which transects a body of mineralisation. These methods can map out sulphide bodies within the earth in 3 dimensions, and provide information to geologists to direct further exploratory drilling on known mineralisation. Surface loop surveys are rarely used for regional exploration, however in some cases such surveys can be used with success (e.g.; SQUID surveys for nickel ore bodies).

Electric-resistance methods such as induced polarization methods can be useful for directly detecting sulfide bodies, coal and resistive rocks such as salt and carbonates.

Hydrocarbon Exploration

Seismic reflection techniques are the most widely used geophysical technique in hydrocarbon exploration. They are used to map the subsurface distribution of stratigraphy and its structure which can be used to delineate potential hydrocarbon accumulations. Well logging is another widely used technique as it provides necessary high resolution information about rock and fluid properties in a vertical section, although they are limited in areal extent. This limitation in areal extent is the reason why seismic reflection techniques are so popular; they provide a method for interpolating and extrapolating well log information over a much larger area.

Gravity and magnetics are also used, with considerable frequency, in oil and gas exploration. These can be used to determine the geometry and depth of covered geological structures including uplifts, subsiding basins, faults, folds, igneous intrusions and salt diapirs due to their unique density and magnetic susceptibility signatures compared to the surrounding rocks.

Remote sensing techniques, specifically hyperspectral imaging, have been used to detect hydrocarbon microseepages using the spectral signature of geochemically altered soils and vegetation.

Magnetotellurics and Controlled source electro-magnetics can provide pseudo-direct detection of hydrocarbons by detecting resistivity changes. It can also complement seismic data when imaging below salt.

Civil Engineering

Ground Penetrating Radar

Ground penetrating radar is a non-invasive technique, and is used within civil construction and engineering for a variety of uses, including detection of abilities (buried water, gas, sewerage, electrical and telecommunication cables), mapping of soft soils and overburden for geotechnical characterization, and other similar uses.

Spectral-analysis-of-surface-waves

The Spectral-Analysis-of-Surface-Waves (SASW) method is another non-invasive technique, which is widely used in practice to image the shear wave velocity profile of the soil. The SASW method relies on the dispersive nature of Raleigh waves in layered media, i.e., the wave-velocity depends on the load's frequency. A material profile, based on the SASW method, is thus obtained according to: a) constructing an experimental dispersion curve, by performing field experiments, each time using a different loading frequency, and measuring the surface wave-speed for each frequency; b) constructing a theoretical dispersion curve, by assuming a trial distribution for the material properties of a layered profile; c) varying the material properties of the layered profile, and repeating the previous step, until a match between the experimental dispersion curve, and the theoretical dispersion curve is attained. The SASW method renders a layered (one-dimensional) shear wave velocity profile for the soil.

Full Waveform Inversion

Full-waveform-inversion (FWI) methods are among the most recent techniques for geotechnical site characterization, and are still under continuous development. The method is fairly general, and is capable of imaging the arbitrarily heterogeneous compressional and shear wave velocity profiles of the soil.

Elastic waves are used to probe the site under investigation, by placing seismic vibrators on the ground surface. These waves propagate through the soil, and due to the heterogeneous geological structure of the site under investigation, multiple reflections and refractions occur. The response of the site to the seismic vibrator is measured by sensors (geophones), also placed on the ground surface. Two key-components are required for the profiling based on full-waveform inversion. These components are: a) a computer model for the simulation of elasic waves in semi-infinite domains; and b) an optimization framework, through which the computed response is matched to the measured response, via iteratively updating an initially assumed material distribution for the soil.

Other Techniques

Civil engineering can also use remote sensing information for topographical mapping, planning and environmental impact assessment. Airborne electromagnetic surveys are also used to characterize soft sediments in planning and engineering roads, dams and other structures.

Magnetotellurics has proven useful for delineating groundwater reservoirs, mapping faults around areas where hazardous substances are stored (e.g. nuclear power stations and nuclear waste storage facilities), and earthquake precursor monitoring in areas with major structures such as hydro-electric dams subject to high levels of seismic activity.

BS 5930 is the standard used in the UK as a code of practice for site investigations.

Archaeology

Ground penetrating radar can be used to map buried artifacts, such as graves, mortuaries, wreck sites, and other shallowly buried archaeological sites.

Ground magnetometric surveys can be used for detecting buried ferrous metals, useful in surveying shipwrecks, modern battlefields strewn with metal debris, and even subtle disturbances such as large-scale ancient ruins.

Sonar systems can be used to detect shipwrecks.

Forensics

Ground penetrating radar can be used to detect grave sites.

Unexploded Ordnance Detection

Magnetic and electromagnetic surveys can be used to locate unexploded ordnance.

Seismic Refraction

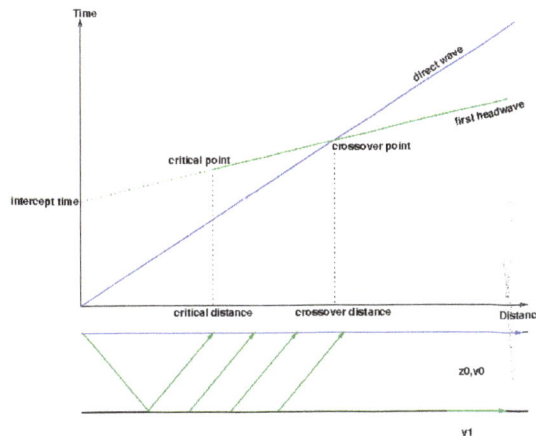

Propagating seismic waves (bottom) and related travel time diagram (top) of the direct (blue)
and the first refracted phase (green)

seismic refraction is a geophysical principle governed by Snell's Law. Used in the fields of engineering geology, geotechnical engineering and exploration geophysics, seismic refraction traverses (seismic lines) are performed using a seismograph(s) and/or geophone(s), in an array and an energy source. The seismic refraction method utilizes the refraction of seismic waves on geologic layers and rock/soil units in order to characterize the subsurface geologic conditions and geologic structure.

The methods depend on the fact that seismic waves have differing velocities in different types of soil (or rock): in addition, the waves are refracted when they cross the boundary between different types (or conditions) of soil or rock. The methods enable the general soil types and the approximate depth to strata boundaries, or to bedrock, to be determined.

P-Wave Refraction (Aka Compression Wave Refraction)

P-wave refraction evaluates the compression wave generated by the seismic source located at a known distance from the array. The wave is generated by vertically striking a striker plate with a sledgehammer, shooting a seismic shotgun into the ground, or detonating an explosive charge in the ground. Since the compression wave is the fastest of the seismic waves, it is sometimes referred to as the primary wave and is usually more-readily identifiable within the seismic recording as compared to the other seismic waves.

S-Wave Refraction (Aka Shear Wave Refraction)

S-wave refraction evaluates the shear wave generated by the seismic source located at a known distance from the array. The wave is generated by horizontally striking an object on the ground surface to induce the shear wave. Since the shear wave is the second fastest wave, it is sometimes referred to as the secondary wave. When compared to the compression wave, the shear wave is approximately one-half (but may vary significantly from this estimate) the velocity depending on the medium.

Seismic Tomography

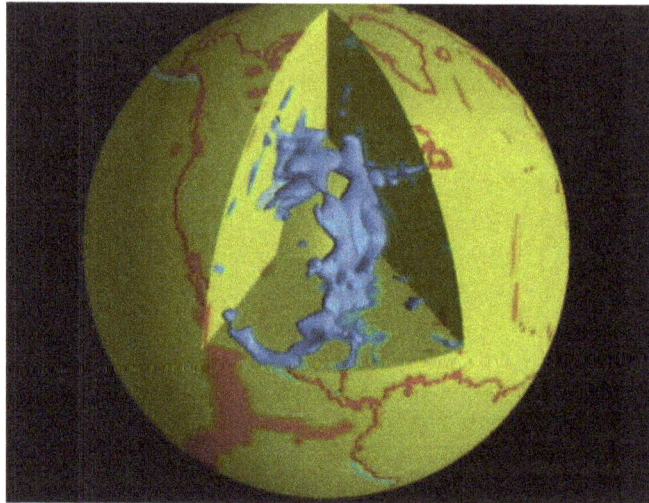

NASA tomographic image of the subducted Farallon Plate in the mantle beneath eastern North America.

Seismic tomography is a technique for imaging the subsurface of the Earth with seismic waves produced by earthquakes or explosions. P-, S-, and surface waves can be used for tomographic models of different resolutions based on seismic wavelength, wave source distance, and the seismograph array coverage. The data received at seismometers are used to solve an inverse problem, wherein the locations of reflection and refraction of the wave paths are determined. This solution can be used to create 3D images of velocity anomalies which may be interpreted as structural, thermal, or compositional variations. Geoscientists use these images to better understand core, mantle, and plate tectonic processes.

Theory

Tomography is solved as an inverse problem. Seismic travel time data are compared to an initial Earth model and the model is modified until the best possible fit between the model predictions and observed data is found. Seismic waves would travel in straight lines if Earth was of uniform composition, but the compositional layering, tectonic structure, and thermal variations reflect and refract seismic waves. The location and magnitude of these variations can be calculated by the inversion process, although solutions to tomographic inversions are non-unique.

Seismic tomography is similar to medical x-ray computed tomography (CT scan) in that a computer processes receiver data to produce a 3D image, although CT scans use attenuation instead of traveltime difference. Seismic tomography has to deal with the analysis of curved ray paths which are reflected and refracted within the earth and potential uncertainty in the location of the earth-quake hypocenter. CT scans use linear x-rays and a known source.

History

Seismic tomography requires large datasets of seismograms and well-located earthquake or explosion sources. These became more widely available in the 1960s with the expansion of global seismic networks and in the 1970s when digital seismograph data archives were established. These developments occurred concurrently with advancements in computing power that were required to solve inverse problems and generate theoretical seismograms for model testing.

In 1977, P-wave delay times were used to create the first seismic array-scale 2D map of seismic velocity. In the same year, P-wave data were used to determine 150 spherical harmonic coefficients for velocity anomalies in the mantle. The first model using iterative techniques, required when there are a large numbers of unknowns, was done in 1984. This built upon the first radially aniso-tropic model of the Earth, which provided the required initial reference frame to compare tomo-graphic models to for iteration. Initial models had resolution of ~3000 to 5000 km, as compared to the few hundred kilometer resolution of current models.

Seismic tomographic models improve with advancements in computing and expansion of seismic networks. Recent models of global body waves used over 10^7 traveltimes to model 10^5 to 10^6 unknowns.

Process

Seismic tomography uses seismic records to create 2D and 3D images of subsurface anomalies by solving large inverse problems such that generate models consistent with observed data. Various methods are used to resolve anomalies in the crust and lithosphere, shallow mantle, whole mantle, and core based on the availability of data and types of seismic waves that penetrate the region at a suitable wavelength for feature resolution. The accuracy of the model is limited by availability and accuracy of seismic data, wave type utilized, and assumptions made in the model.

P-wave data are used in most local models and global models in areas with sufficient earthquake and seismograph density. S- and surface wave data are used in global models when this coverage is not sufficient, such as in ocean basins and away from subduction zones. First-arrival times are the most widely used, but models utilizing reflected and refracted phases are used in more complex models, such as those imaging the core. Differential traveltimes between wave phases or types are also used.

Local Tomography

Local tomographic models are often based on a temporary seismic array targeting specific areas, unless in a seismically active region with extensive permanent network coverage. These allow for the imaging of the crust and upper mantle.

- *Diffraction and wave equation tomography* use the full waveform, rather than just the first arrival times. The inversion of amplitude and phases of all arrivals provide more de-tailed density information than transmission traveltime alone. Despite the theoretical appeal, these methods are not widely employed because of the computing expense and difficult inversions.

- *Reflection tomography* originated with exploration geophysics. It uses an artificial source to resolve small-scale features at crustal depths. *Wide-angle tomography* is similar, but with a wide source to receiver offset. This allows for the detection of seismic waves refracted from sub-crustal depths and can determine continental architecture and details of plate margins. These two methods are often used together.

- *Local earthquake tomography* is used in seismically active regions with sufficient seismometer coverage. Given the proximity between source and receivers, a precise earthquake focus location must be known. This requires the simultaneous iteration of both structure and focus locations in model calculations.

- *Teleseismic tomography* uses waves from distant earthquakes that deflect upwards to a local seismic array. The models can reach depths similar to the array aperture, typically to depths for imaging the crust and lithosphere (a few hundred kilometers). The waves travel near 30° from vertical, creating a vertical distortion to compact features.

Regional or Global Tomography

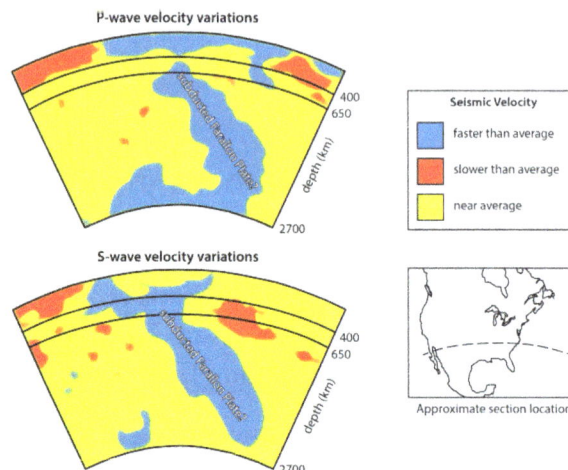

Simplified and interpreted P- and S-wave velocity variations in the mantle across southern North America showing the subducted Farallon Plate.

Regional to global scale tomographic models are generally based on long wavelengths. Various models have better agreement with each other than local models due to the large feature size they image, such as subducted slabs and superplumes. The trade off from whole mantle to whole earth coverage is the coarse resolution (hundreds of kilometers) and difficulty imaging small features (e.g. narrow plumes). Although often used to image different parts of the subsurface, P- and S-wave derived models broadly agree where there is image overlap. These models use data from both permanent seismic stations and supplementary temporary arrays.

- First arrival traveltime *P-wave* data are used to generate the highest resolution tomographic images of the mantle. These models are limited to regions with sufficient seismograph coverage and earthquake density, therefore cannot be used for areas such as inactive plate interiors and ocean basins without seismic networks. Other phases of P-waves are used to image the deeper mantle and core.

- In areas with limited seismograph or earthquake coverage, multiple phases of *S-waves* can be used for tomographic models. These are of lower resolution than P-wave models, due to the distances involved and fewer bounce-phase data available. S-waves can also be used in conjunction with P-waves for differential arrival time models.

- *Surface waves* can be used for tomography of the crust and upper mantle where no body wave (P and S) data are available. Both Rayleigh and Love waves can be used. The low frequency waves lead to low resolution models, therefore these models have difficulty with crustal structure. *Free oscillations*, or normal mode seismology, are the long wavelength, low frequency movements of the surface of the earth which can be thought of as a type of surface wave. The frequencies of these oscillations can be obtained through Fourier transformation of seismic data. The models based on this method are of broad scale, but have the advantage of relatively uniform data coverage as compared to data sourced directly from earthquakes.

- *Attenuation tomography* attempts to extract the anaelastic signal from the elastic-dominated waveform of seismic waves. The advantage of this method is its sensitivity to temperature, thus ability to image thermal features such as mantle plumes and subduction zones. Both surface and body waves have been used in this approach.

- *Ambient noise tomography* cross-correlates waveforms from random wavefields generated by oceanic and atmospheric disturbances which subsequently become diffuse within the Earth. This method has produced high-resolution images and is an area of active research.

- Waveforms are modeled as rays in seismic analysis, but all waves are affected by the material near the ray path. The finite frequency effect is the result the surrounding medium has on a seismic record. *Finite frequency tomography* accounts for this in determining both travel time and amplitude anomalies, increasing image resolution. This has the ability to resolve much larger variations (i.e. 10-30%) in material properties.

Applications

Seismic tomography can resolve anisotropy, anelasticity, density, and bulk sound density. Variations in these parameters may be a result of thermal or chemical differences, which are attributed to processes such as mantle plumes, subducting slabs, and mineral phase changes. Larger scale features that can be imaged with tomography include the high velocities beneath continental shields and low velocities under ocean spreading centers.

Hotspots

The mantle plume hypothesis proposes that areas of volcanism not readily explained by plate tectonics, called hotspots, are a result of thermal upwelling from as deep as the core-mantle boundary that become diapirs in the crust. This is an actively contested theory, although tomographic images suggest there are anomalies beneath some hotspots. The best imaged of these are large low-shear-velocity provinces, or superplumes, visible on S-wave models of the lower mantle and believed to reflect both thermal and compositional differences.

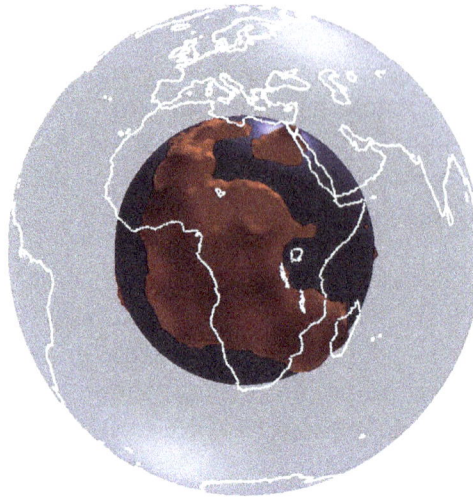

Tomographic image of the African large low-shear-velocity province (superplume).

The Yellowstone hotspot is responsible for volcanism at the Yellowstone Caldera and a series of extinct calderas along the Snake River Plain. The Yellowstone Geodynamic Project sought to image the plume beneath the hotspot. They found a strong low-velocity body from ~30 to 250 km depth beneath Yellowstone and a weaker anomaly from 250 to 650 km depth which dipped 60° west-northwest. The authors attribute these features to the mantle plume beneath the hotspot being deflected eastward by flow in the upper mantle seen in S-wave models.

The Hawaii hotspot lies beneath the Hawaiian Islands and Emperor Seamounts. Tomographic images show it to be 500 to 600 km wide and up to 2,000 km deep.

Subduction Zones

Subducting plates are colder than the mantle into which they are moving. This creates a fast anomaly that is visible in tomographic images. Both the Farallon plate that subducted beneath the west coast of North America and the northern portion of the Indian plate that has subducted beneath Asia have been imaged with tomography.

Limitations

Global seismic networks have expanded steadily since the 1960s, but are still concentrated on continents and in seismically active regions. Oceans, particularly in the southern hemisphere, are under-covered. Tomographic models in these areas will improve when more data becomes available. The uneven distribution of earthquakes naturally biases models to better resolution in seismically active regions.

The type of wave used in a model limits the resolution it can achieve. Longer wavelengths are able to penetrate deeper into the earth, but can only be used to resolve large features. Finer resolution can be achieved with surface waves, with the trade off that they cannot be used in models of the deep mantle. The disparity between wavelength and feature scale causes anomalies to appear of reduced magnitude and size in images. P- and S-wave models respond differently to the types of anomalies depending on the driving material property. First arrival time based models naturally prefer faster pathways, causing models based on these data to have lower resolution of slow (often hot) features. Shallow models must also consider the significant lateral velocity variations in continental crust.

Seismic tomography provides only the current velocity anomalies. Any prior structures are unknown and the slow rates of movement in the subsurface (mm to cm per year) prohibit resolution of changes over modern timescales.

Tomographic solutions are non-unique. Although statistical methods can be used to analyze the validity of a model, unresolvable uncertainty remains. This contributes to difficulty comparing the validity of different model results.

Computing power limits the amount of seismic data, number of unknowns, mesh size, and iterations in tomographic models. This is of particular importance in ocean basins, which due to limited network coverage and earthquake density require more complex processing of distant data. Shallow oceanic models also require smaller model mesh size due to the thinner crust.

Tomographic images are typically presented with a color ramp representing the strength of the anomalies. This has the consequence of making equal changes appear of differing magnitude based on visual perceptions of color, such as the change from orange to red being more subtle than blue to yellow. The degree of color saturation can also visually skew interpretations. These factors should be considered when analyzing images.

Gravity Gradiometry

Gravity gradiometry is the study and measurement of variations in the acceleration due to gravity. The gravity gradient is the spatial rate of change of gravitational acceleration.

Gravity gradiometry is used by oil and mineral prospectors to measure the density of the subsurface, effectively the rate of change of rock properties. From this information it is possible to build a picture of subsurface anomalies which can then be used to more accurately target oil, gas and mineral deposits. It is also used to image water column density, when locating submerged objects, or determining water depth (bathymetry). Physical scientists use gravimeters to determine the exact size and shape of the earth and they contribute to the gravity compensations applied to inertial navigation systems.

Units

The unit of gravity gradient is the eotvos (abbreviated as E), which is equivalent to 10^{-9} s^{-2} (or 10^{-4} mGal/m). A person walking past at a distance of 2 metres would provide a gravity gradient signal approximately one E. Mountains can give signals of several hundred Eotvos.

Gravity Gradient Tensor

Full tensor gradiometers measure the rate of change of the gravity vector in all three perpendicular directions giving rise to a gravity gradient tensor (Fig 1).

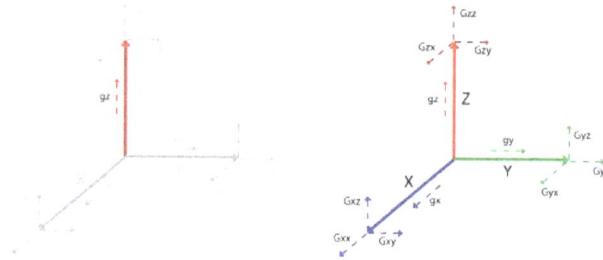

Fig 1. Conventional gravity measures ONE component of the gravity field in the vertical direction Gz (LHS), Full tensor gravity gradiometry measures ALL components of the gravity field (RHS)

Comparison to Gravity

Being the derivatives of gravity, the spectral power of gravity gradient signals is pushed to higher frequencies. This generally makes the gravity gradient anomaly more localised to the source than the gravity anomaly. The table (below) and graph (Fig 2) compare the g_z and G_{zz} responses from a point source,

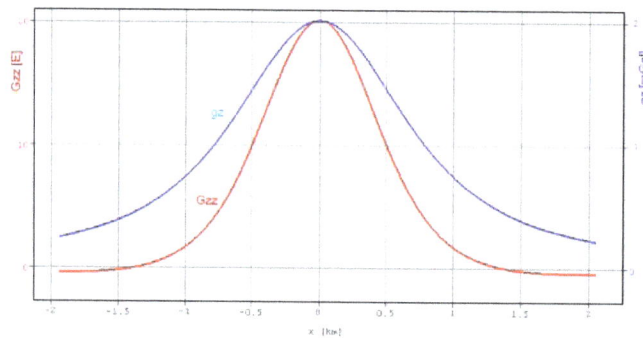

Fig 2. Vertical gravity and gravity gradient signals from a point source buried at 1 km depth

Conversely, gravity measurements have more signal power at low frequency therefore making them more sensitive to regional signals and deeper sources.

Dynamic Survey Environments (Airborne and Marine)

The derivative measurement sacrifices the overall energy in the signal, but significantly reduces the noise due to motional disturbance. On a moving platform, the acceleration disturbance measured by the two accelerometers is the same so that when forming the difference, it cancels in the gravity gradient measurement. This is the principal reason for deploying gradiometers in airborne and marine surveys where the acceleration levels are orders of magnitude greater than the signals of interest. The signal to noise ratio benefits most at high frequency (above 0.01 Hz), where the airborne acceleration noise is largest.

Applications

Gravity gradiometry has predominately been used to image subsurface geology to aid hydrocarbon and mineral exploration. Over 2.5 million line km has now been surveyed using the technique. The surveys highlight gravity anomalies that can be related to geological features such as Salt diapirs,

Fault systems, Reef structures, Kimberlite pipes, etc. Other applications include tunnel and bunker detection and the recent GOCE mission that aims to improve the knowledge of ocean circulation.

Gravity Gradiometers

Lockheed Martin Gravity Gradiometers

During the 1970s, as an executive in the Dept. of Defense, John Brett initiated the development of the gravity gradiometer to support the Trident 2 system. A committee was commissioned to seek commercial applications for the Full Tensor Gradient (FTG) system that was developed by Bell Aerospace (later acquired by Lockheed Martin) and was being deployed on US Navy *Ohio*-class Trident submarines designed to aid covert navigation. As the Cold War came to a close, the US Navy released the classified technology and opened the door for full commercialization of the technology. The existence of the gravity gradiometer was famously exposed in the film *The Hunt for Red October* released in 1990.

There are two types of Lockheed Martin gravity gradiometers currently in operation: the 3D Full Tensor Gravity Gradiometer (FTG; deployed in either a fixed wing aircraft or a ship) and the FALCON gradiometer (a partial tensor system with 8 accelerometers and deployed in a fixed wing aircraft or a helicopter). The 3D FTG system contains three gravity gradiometry instruments (GGIs), each consisting of two opposing pairs of accelerometers arranged on a spinning disc with measurement direction in the spin direction.

Other Gravity Gradiometers

Electrostatic gravity gradiometer

> This is the gravity gradiometer deployed on the European Space Agency's GOCE mission. It is a three-axis diagonal gradiometer based on three pairs of electrostatic servo-controlled accelerometers.

ARKeX Exploration Gravity Gradiometer

> An evolution of technology originally developed for European Space Agency, the Exploration Gravity Gradiometer (EGG), developed by ARKeX, uses two key principles of superconductivity to deliver its performance: the Meissner effect, which provides levitation of the EGG proof masses and flux quantization, which gives the EGG its inherent stability. The EGG has been specifically designed for high dynamic survey environments.

Ribbon Sensor Gradiometer

> The Gravitec gravity gradiometer sensor consists of a single sensing element (a ribbon) that responds to gravity gradient forces. It is designed for borehole applications.

UWA Gravity Gradiometer

> The University of Western Australia (aka VK-1) Gravity Gradiometer is a superconducting instrument which uses an orthogonal quadrupole responder (OQR) design based on pairs of micro-flexure supported balance beams.

Gedex Gravity Gradiometer

The Gedex gravity gradiometer (AKA High-Definition Airborne Gravity Gradiometer, HD-AGG) is also a superconducting OQR-type gravity gradiometer, based on technology developed at the University of Maryland.

Aeromagnetic Survey

An aeromagnetic survey is a common type of geophysical survey carried out using a magnetometer aboard or towed behind an aircraft. The principle is similar to a magnetic survey carried out with a hand-held magnetometer, but allows much larger areas of the Earth's surface to be covered quickly for regional reconnaissance. The aircraft typically flies in a grid-like pattern with height and line spacing determining the resolution of the data (and cost of the survey per unit area).

Method

As the aircraft flies, the magnetometer measures and records the total intensity of the magnetic field at the sensor, which is a combination of the desired magnetic field generated in the Earth as well as tiny variations due to the temporal effects of the constantly varying solar wind and the magnetic field of the survey aircraft. By subtracting the solar, regional, and aircraft effects, the resulting aeromagnetic map shows the spatial distribution and relative abundance of magnetic minerals (most commonly the iron oxide mineral magnetite) in the upper levels of the Earth's crust. Because different rock types differ in their content of magnetic minerals, the magnetic map allows a visualization of the geological structure of the upper crust in the subsurface, particularly the spatial geometry of bodies of rock and the presence of faults and folds. This is particularly useful where bedrock is obscured by surface sand, soil or water. Aeromagnetic data was once presented as contour plots, but now is more commonly expressed as thematic (colored) and shaded computer generated pseudo-topography images. The apparent hills, ridges and valleys are referred to as aeromagnetic anomalies. A geophysicist can use mathematical modeling to infer the shape, depth and properties of the rock bodies responsible for the anomalies.

Airplanes are normally used for high-level reconnaissance surveys in gentle terrain, and helicopters are used in mountainous terrain or where more detail is required. UAVs can also be used in aeromagnetic survey work.

Uses

Aeromagnetic surveys are widely used to aid in the production of geological maps and are also commonly used during mineral exploration and petroleum exploration. Some mineral deposits are associated with an increase or decrease in the abundance of magnetic minerals, and occasionally the sought after commodity may itself be magnetic (e.g. iron ore deposits), but often the elucidation of the subsurface structure of the upper crust is the most valuable contribution of the aeromagnetic data.

This helicopter is equipped with a magnetometer array. It flies six feet above ground at speeds of 30 to 40 mph.

Aeromagnetic surveys were performed in World War II to detect submarines using a Magnetic Anomaly Detector attached to an aircraft. Aeromagnetic surveys are also used to perform reconnaissance mapping of unexploded ordnance (UXO). The aircraft is typically a helicopter, as the sensors must be close to the ground (relative to mineral exploration) to be effective. Electromagnetic methods are also used for this purpose.

Eurocopter AS350 geophysical survey helicopter equipped with an aeromagnetic survey system.

Electrical Resistivity Tomography

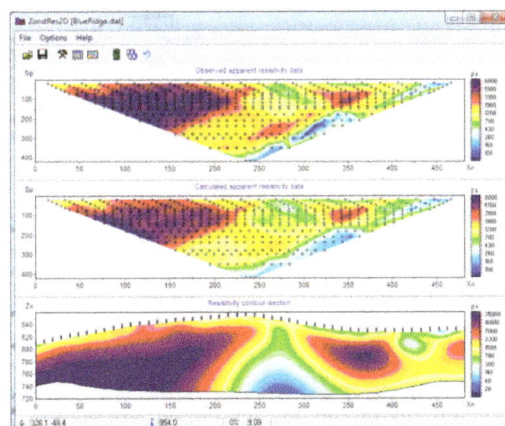

2D resistivity inversion of ERT data

Deployment of a permanent electrical resistivity tomography profile on a longitudinal section of an active landslide.

Electrical resistivity tomography (ERT) or electrical resistivity imaging (ERI) is a geophysical technique for imaging sub-surface structures from electrical resistivity measurements made at the surface, or by electrodes in one or more boreholes. If the electrodes are suspended in the boreholes, deeper sections can be investigated. It is closely related to the medical imaging technique electrical impedance tomography (EIT), and mathematically is the same inverse problem. In contrast to medical EIT however ERT is essentially a direct current method.

A related geophysical method, induced polarization, measures the transient response. The technique evolved from techniques of electrical prospecting that predate digital computers, where layers or anomalies were sought rather than images. Early work on the mathematical problem in the 1930s assumed a layered medium (for example Langer, Slichter). Tikhonov who is best known for his work on regularization of inverse problems also worked on this problem. He explains in detail how to solve the ERT problem in a simple case of 2-layered medium. During the 1940s he collaborated with geophysicists and without the aid of computers they discovered large deposits of copper. As a result, they were awarded a State Prize of Soviet Union.

When adequate computers became widely available the inverse problem of ERT could be solved numerically, and the work of Loke and Barker at Birmingham University was among the first such solution, and their approach is still widely used.

With the advancement in the field of Electrical Resistivity Tomography (ERT) from 1D to 2D and now-a- days 3D, ERT has explored many fields. The applications of ERT include fault investigation, ground water table investigation, soil moisture content determination and many others. In industrial process imaging ERT can be used in a similar fashion to medical EIT, to image the distribution of conductivity in mixing vessels and pipes. In this context it is usually called *Electrical Resistance Tomography*, emphasising the quantity that is measured rather than imaged.

Magnetotellurics

Magnetotellurics (MT) is an electromagnetic geophysical method for inferring the earth's subsurface electrical conductivity from measurements of natural geomagnetic and geoelectric field variation at the Earth's surface. Investigation depth ranges from 300 m below ground by recording higher frequencies down to 10,000 m or deeper with long-period soundings. Developed in the USSR and France during the 1950s, MT is now an international academic discipline and is used in exploration surveys around the world. Commercial uses include hydrocarbon (oil and gas) exploration, geothermal exploration, mining exploration, as well as hydrocarbon and groundwater monitoring. Research applications include experimentation to further develop the MT technique, long-period deep crustal exploration, and earthquake precursor prediction research.

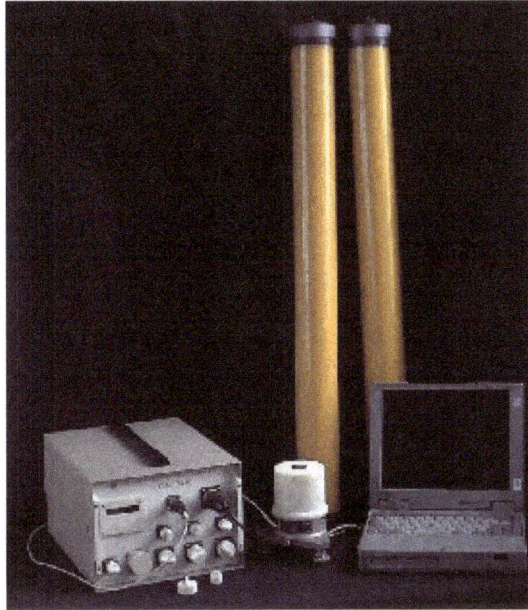
Magnetotelluric station

History

The magnetotelluric technique was introduced by the French geophysicist Louis Cagniard and Russian geophysicist Andrey Nikolayevich Tikhonov in the early 1950s. With advances in instrumentation, processing and modeling, MT has become one of the most important tools in deep Earth research.

Since first being created in the 1950s, magnetotelluric sensors, receivers and data processing techniques have followed the general trends in electronics, becoming less expensive and more capable with each generation. Major advances in MT instrumentation and technique include the shift from analog to digital hardware, the advent of remote referencing, GPS time-based synchronization, and 3D data acquisition and processing.

Commercial Applications

Hydrocarbon Exploration

For hydrocarbon exploration, MT is mainly used as a complement to the primary technique of reflection seismology exploration. While seismic imaging is able to image subsurface structure, it cannot detect the changes in resistivity associated with hydrocarbons and hydrocarbon-bearing formations. MT does detect resistivity variations in subsurface structures, which can differentiate between structures bearing hydrocarbons and those that do not.

At a basic level of interpretation, resistivity is correlated with different rock types. High-velocity layers are typically highly resistive, whereas sediments – porous and permeable – are typically much less resistive. While high-velocity layers are an acoustic barrier and make seismic ineffective, their electrical resistivity means the magnetic signal passes through almost unimpeded. This allows MT to see deep beneath these acoustic barrier layers, complementing the seismic data and assisting interpretation. 3-D MT survey results in Uzbekistan (32 x 32 grid of soundings) have guided further seismic mapping of a large known gas-bearing formation with complex subsurface geology.

China National Petroleum Corporation (CNPC) uses onshore MT more than any other oil company in the world, conducting thousands of MT soundings for hydrocarbon exploration and mapping throughout all of China.

Mining Exploration

MT is used for various base metals (nickel and precious metals exploration, as well as for kimberlite) mapping.

INCO's 1991 proof-of-concept study in Sudbury, Ontario, Canada sensed a 1750-meter-deep nickel deposit. Falconbridge followed with a feasibility study in 1996 that accurately located two Ni-Cu mineralized zones at about 800 m and 1350 m depth. Since then, both major and junior mining companies are increasingly using MT and AMT for both brownfields and greenfields exploration. Significant MT mapping work has been done on areas of the Canadian Shield.

Diamond exploration, by detecting kimberlites, is also a proven application.

Geothermal Exploration

MT geothermal exploration measurements allow detection of resistivity anomalies associated with productive geothermal structures, including faults and the presence of a cap rock, and allow for estimation of geothermal reservoir temperatures at various depths. Dozens of MT geothermal exploration surveys have been completed in Japan and the Philippines since the early 1980s, helping to identify several hundred megawatts of renewable power at places such as the Hatchobaru plant on Kyushu and the Togonang plant on Leyte. Geothermal exploration with MT has also been done in the United States, Iceland, New Zealand, Hungary, China, Ethiopia, Indonesia, Peru, Australia, and India.

Other Commercial Applications

MT is also used for groundwater exploration and mapping, hydrocarbon reservoir monitoring, deep investigation (100 km) of the electrical properties of the bedrock for high-voltage direct current (HVDC) transmission systems, carbon dioxide sequestration, and other environmental engineering applications (e.g. nuclear blast site monitoring and nuclear waste disposal site monitoring).

Research Applications

Crustal Research

MT has been used to investigate the distribution of silicate melts in the Earth's mantle and crust; large investigations have focused on the continental US (National Science Foundation EarthScope MT Program), the East Pacific Rise and the Tibetan Plateau. Other research work aims to better understand the plate-tectonic processes in the highly complex three-dimensional region formed by the collision of the African and European plates.

Earthquake Precursor Prediction Research

Fluctuations in the MT signal may be able to predict the onset of seismic events. Stationary MT monitoring systems have been installed in Japan since April 1996, providing a continuous recording of MT signals at the Wakuya Station (previously at the Mizusawa Geodetic Observatory) and the Esashi Station of the Geographical Survey Institute of Japan (GSIJ). These stations measure fluctuations in the Earth's electromagnetic field that correspond with seismic activity. The raw geophysical time-series data from these monitoring stations is freely available to the scientific community, enabling further study of the interaction between electromagnetic events and earthquake activity. The MT time series data from the GSIJ earthquake monitoring stations is available online at http://vldb.gsi.go.jp/sokuchi/geomag/menu_03/mt_data-e.html

Additional MT earthquake precursor monitoring stations in Japan are located in Kagoshima, in Sawauchi, and on Shikoku. Similar stations are also deployed in Taiwan on Penghu Island, as well as in the Fushan Reserve on the island of Taiwan proper.

POLARIS is a Canadian research program investigating the structure and dynamics of the Earth's lithosphere and the prediction of earthquake ground motion.

Theory and Practice

Energy Sources

Solar energy and lightning cause natural variations in the earth's magnetic field, inducing electric currents (known as telluric currents) under the Earth's surface. Simultaneous measurements of orthogonal components of the electric and magnetic fields are recorded, with the results calculated as the impedance tensor. A subsurface resistivity model is then created using this tensor.

Different rocks, sediments and geological structures have a wide range of different electrical conductivities. Measuring electrical resistivity allows different materials and structures to be distinguished from one another and can improve knowledge of tectonic processes and geologic structures.

The Earth's naturally varying electric and magnetic fields are measured over a wide range of magnetotelluric frequencies from 10,000 Hz to 0.0001 Hz (10,000 s). These fields are due to electric currents flowing in the Earth and the magnetic fields that induce these currents. The magnetic fields are produced mainly by the interaction between the solar wind and the magnetosphere.

In addition, worldwide thunderstorm activity causes magnetic fields at frequencies above 1 Hz. Combined, these natural phenomena create strong MT source signals over the entire frequency spectrum.

The ratio of the electric field to magnetic field provides simple information about subsurface conductivity. Because the skin effect phenomenon affects the electromagnetic fields, the ratio at higher frequency ranges gives information on the shallow Earth, whereas deeper information is provided by the low-frequency range. The ratio is usually represented as both apparent resistivity as a function of frequency and phase as a function of frequency.

Depth and Resolution

MT measurements can investigate depths from about 300 m down to hundreds of kilometers, though investigations in the range of 500 m to 10,000 m are typical. Greater depth requires measuring lower frequencies, which in turn requires longer recording times. Very deep, very long-period measurements (mid-crust through upper mantle depths), may require recordings of several days to weeks or more to obtain satisfactory data quality.

Horizontal resolution of MT mainly depends on the distance between sounding locations- closer sounding locations increase the horizontal resolution. Continuous profiling (known as Emap) has been used, with only meters between the edges of each telluric dipole.

Vertical resolution of MT mainly depends on the frequency being measured, as lower frequencies have greater depths of penetration. Accordingly, vertical resolution decreases as depth of investigation increases.

Signal Strength and Recording Times

Magnetic fields in the frequency range of 1 Hz to approximately 20 kHz are part of the audio-magnetotelluric (AMT) range. These are parallel to the Earth surface and move towards the Earth's centre. This large frequency band allows for a range of depth penetration from several metres to several kilometres below the Earth's surface. Due to the nature of magnetotelluric source, the waves generally fluctuate in amplitude height. Long recording times are needed to ascertain usable reading due to the fluctuations and the low signal strength. Generally, the signal is weak between 1 and 5 kHz, which is a crucial range in detecting the top 100 m of geology. The magnetotelluric method is also used in marine environments for hydrocarbon exploration and lithospheric studies. Due to the screening effect of the electrically conductive sea water, a usable upper limit of the spectrum is around 1 Hz.

2D and 3D Magnetotellurics

Two-dimensional surveys consist of a longitudinal profile of MT soundings over the area of interest, providing two-dimensional "slices" of subsurface resistivity.

Three-dimensional surveys consist of a loose grid pattern of MT soundings over the area of interest, providing a more sophisticated three-dimensional model of subsurface resistivity.

Variants

Audio-magnetotellurics (AMT)

AMT is a higher-frequency magnetotelluric technique for shallower investigations. While AMT has less depth penetration than MT, AMT measurements often take only about one hour to perform (but deep AMT measurements during low-signal strength periods may take up to 24 hours) and use smaller and lighter magnetic sensors. Transient AMT is an AMT variant that records only temporarily during periods of more intense natural signal (transient impulses), improving signal-to-noise-ratio at the expense of strong linear polarization.

Controlled Source Electromagnetics

CSEM Controlled source electro-magnetic is a deep-water offshore variant of controlled source audio magnetotellurics; CSEM is the name used in the offshore oil and gas industry.

Onshore CSEM / CSAMT may be effective where electromagnetic cultural noise (e.g. power lines, electric fences) present interference problems for natural-source geophysical methods. An extensive grounded wire (2 km or more) has currents at a range of frequencies (0.1 Hz to 100 kHz) passed through it. The electric field parallel to the source and the magnetic field which is at right angles are measured. The resistivity is then calculated, and the lower the resistivity, the more likely there is a conductive target (graphite, nickel ore or iron ore). CSAMT is also known in the oil and gas industry as Onshore Controlled Source ElectroMagnetics (Onshore CSEM).

An offshore variant of MT, the marine magnetotelluric (MMT) method, uses instruments and sensors in pressure housings deployed by ship into shallow coastal areas where water is less than 300 m deep. A derivative of MMT is offshore single-channel measurement of the vertical magnetic field only (the Hz, or "tipper"), which eliminates the need for telluric measurements and horizontal magnetic measurements. While the theory is sound, no commercial system is yet available. Furthermore, any such system would require a solution providing for the precise orientation and stabilization of the magnetic sensor.

Exploration Surveys

MT exploration surveys are done to acquire resistivity data which can be interpreted to create a model of the subsurface. Data is acquired at each sounding location for a period of time (overnight soundings are common), with physical spacing between soundings dependant on the target size and geometry, local terrain constraints and financial cost. Reconnaissance surveys can have spacings of several kilometres, while more detailed work can have 200 m spacings, or even adjacent soundings (dipole-to-dipole). MT surveys are carried out year-round in the Americas, Europe, Asia, Africa and Oceania.

The HSE impact of MT exploration is relatively low because of light-weight equipment, natural signal sources, and reduced hazards compared to other types of exploration (e.g. no drills, no explosives, and no high currents).

Remote Reference Soundings

Remote Reference is an MT technique used to account for cultural electrical noise by acquiring simultaneous data at more than one MT station. This greatly improves data quality, and may allow acquisition in areas where the natural MT signal is difficult to detect because of man-made EM interference.

Survey Equipment

A typical full suite of MT equipment (for a "five component" sounding) consists of a receiver instrument with five sensors: three magnetic sensors (typically induction coil sensors), and two telluric (electric) sensors. For long-period MT (frequencies below approximately 1–10 Hz), the three discrete magnetic field sensors can typically be replaced with a single compact triaxial fluxgate magnetometer. In many situations, only the telluric sensors will be used, and magnetic data borrowed from other nearby soundings to reduce acquisition costs.

A complete five-component set of MT equipment can be backpack-carried by a small field team (2 to 4 persons) or carried by a light helicopter (such as the MD Helicopters MD 500), allowing deployment in remote and rugged areas. Most MT equipment is capable of reliable operation over a wide range of environmental conditions, with ratings of typically −20 °C to +45 °C, from dry desert to high-humidity (condensing) and partial immersion.

Data Processing and Interpretation

Post-acquisition processing is required to transform raw time-series data into frequency-based inversions. The resulting output of the processing program is used as the input for subsequent interpretation. Processing may include the use of remote reference data or local data only.

Processed MT data is modelled using various techniques to create a subsurface resistivity map, with lower frequencies generally corresponding to greater depth below ground. Anomalies such as faults, hydrocarbons, and conductive mineralization appear as areas of higher or lower resistivity from surrounding structures. For interpretation(inversion) of magnetotelluric data a number of software are used (WinGlink, Geotools MT, ZondMT2D).

Instrument and Sensor Manufacturers

MT instrumentation design and construction is a specialized international activity, with only a small number of companies and scientific organizations having the necessary expertise and technology. Three companies supply most of the commercial-use world market: one in the United States (Zonge International, Inc.), two in Canada (Advanced Geophysical Operations and Services Inc. (AGCOS)) and (Phoenix Geophysics, Ltd.) and one in Germany (Metronix Messgeraete und Elektronik GmbH). Government agencies and smaller companies producing MT instrumentation for internal use include Vega Geophysics, Ltd. in Russia and the Russian Academy of Sciences (SPbF IZMIRAN); and the National Space Research Institute of Ukraine. Geometrics, Inc. of San Jose, California USA produces a high-frequency, hybrid-source AMT system combining both natural fields and a high-frequency (1 to 70 kHz) transmitter, as well as a multi-channel distributed network CSAMT instrument.

Ground-penetrating Radar

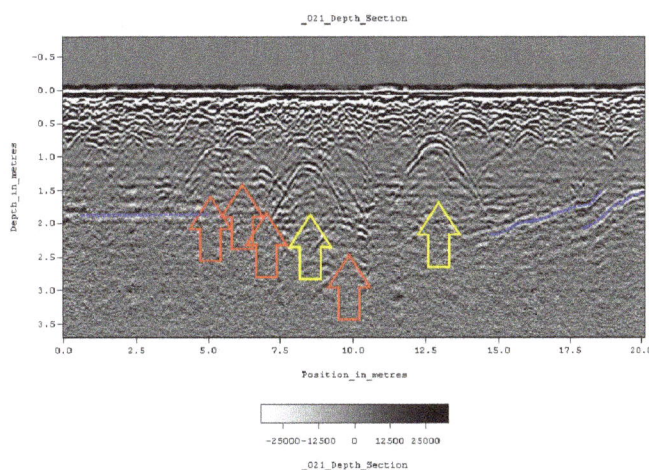

A ground-penetrating radargram collected on a historic cemetery in Alabama, USA. Hyperbolic arrivals (arrows) indicate the presence of diffractors buried beneath the surface, possibly associated with human burials. Reflections from soil layering are also present (dashed lines).

Ground-penetrating radar (GPR) is a geophysical method that uses radar pulses to image the subsurface. This nondestructive method uses electromagnetic radiation in the microwave band (UHF/VHF frequencies) of the radio spectrum, and detects the reflected signals from subsurface structures. GPR can have applications in a variety of media, including rock, soil, ice, fresh water, pavements and structures. In the right conditions, practitioners can use GPR to detect subsurface objects, changes in material properties, and voids and cracks.

GPR uses high-frequency (usually polarized) radio waves, usually in the range 10 MHz to 2.6 GHz. A GPR transmitter emits electromagnetic energy into the ground. When the energy encounters a buried object or a boundary between materials having different permittivities, it may be reflected or refracted or scattered back to the surface. A receiving antenna can then record the variations in the return signal. The principles involved are similar to seismology, except GPR methods implement electromagnetic energy rather than acoustic energy, and energy may be reflected at boundaries where subsurface electrical properties change rather than subsurface mechanical properties as is the case with seismic energy.

The electrical conductivity of the ground, the transmitted center frequency, and the radiated power all may limit the effective depth range of GPR investigation. Increases in electrical conductivity attenuate the introduced electromagnetic wave, and thus the penetration depth decreases. Because of frequency-dependent attenuation mechanisms, higher frequencies do not penetrate as far as lower frequencies. However, higher frequencies may provide improved resolution. Thus operating frequency is always a trade-off between resolution and penetration. Optimal depth of subsurface penetration is achieved in ice where the depth of penetration can achieve several thousand metres (to bedrock in Greenland) at low GPR frequencies. Dry sandy soils or massive dry materials such as granite, limestone, and concrete tend to be resistive rather than conductive, and the depth of penetration could be up to 15-metre (49 ft). In moist and/or clay-laden soils and materials with high electrical conductivity, penetration may be as little as a few centimetres.

Ground-penetrating radar antennas are generally in contact with the ground for the strongest signal strength; however, GPR air-launched antennas can be used above the ground.

Cross borehole GPR has developed within the field of hydrogeophysics to be a valuable means of assessing the presence and amount of soil water.

History

The first patent for a system designed to use continuous-wave radar to locate buried objects was submitted by Gotthelf Leimbach and Heinrich Löwy in 1910, six years after the first patent for radar itself (patent DE 237 944). A patent for a system using radar pulses rather than a continuous wave was filed in 1926 by Dr. Hülsenbeck (DE 489 434), leading to improved depth resolution. A glacier's depth was measured using ground penetrating radar in 1929 by W. Stern.

Further developments in the field remained sparse until the 1970s, when military applications began driving research. Commercial applications followed and the first affordable consumer equipment was sold in 1985.

Applications

Ground penetrating radar in use near Stillwater, Oklahoma, USA in 2010

GPR has many applications in a number of fields. In the Earth sciences it is used to study bedrock, soils, groundwater, and ice. It is of some utility in prospecting for gold nuggets and for diamonds in alluvial gravel beds, by finding natural traps in buried stream beds that have the potential for accumulating heavier particles. The Chinese lunar rover Yutu has a GPR on its underside to investigate the soil and crust of the Moon.

Engineering applications include nondestructive testing (NDT) of structures and pavements, locating buried structures and utility lines, and studying soils and bedrock. In environmental remediation, GPR is used to define landfills, contaminant plumes, and other remediation sites, while in archaeology it is used for mapping archaeological features and cemeteries. GPR is used in law enforcement for locating clandestine graves and buried evidence. Military uses include detection of mines, unexploded ordnance, and tunnels.

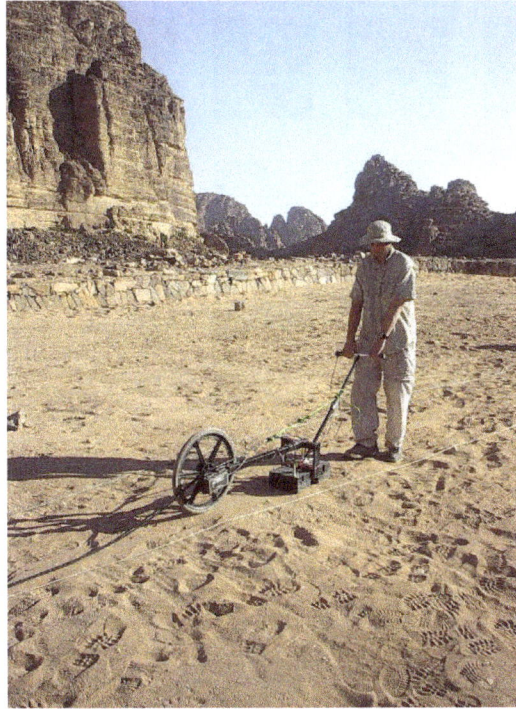

Ground penetrating radar survey of an archaeological site in Jordan

Borehole radars utilizing GPR are used to map the structures from a borehole in underground mining applications. Modern directional borehole radar systems are able to produce three-dimensional images from measurements in a single borehole.

One of the other main applications for ground-penetrating radars is for locating underground utilities. Standard electromagnetic induction utility locating tools require utilities to be conductive. These tools are ineffective for locating plastic conduits or concrete storm and sanitary sewers. Since GPR detects variations in dielectric properties in the subsurface, it can be highly effective for locating non-conductive utilities.

GPR is often used on the Channel 4 television programme *Time Team* which uses the technology to determine a suitable area for examination by means of excavations. In 1992 GPR was used to recover £150,000 in cash that kidnapper Michael Sams received as a ransom for an estate agent he had kidnapped after Sams buried the money in a field.

Archaeology

Ground penetrating radar survey is one method used in archaeological geophysics. GPR can be used to detect and map subsurface archaeological artifacts, features, and patterning.

The concept of radar is familiar to most people. With ground penetrating radar, the radar signal – an electromagnetic pulse – is directed into the ground. Subsurface objects and stratigraphy (layering) will cause reflections that are picked up by a receiver. The travel time of the reflected signal indicates the depth. Data may be plotted as profiles, as planview maps isolating specific depths, or as three-dimensional models.

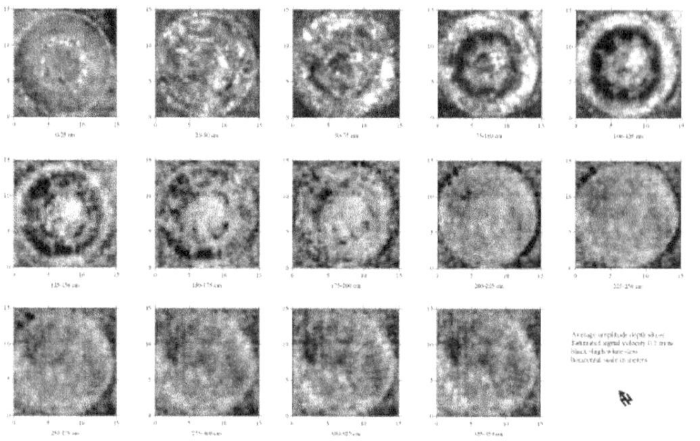

GPR depth slices showing a crypt in a historic cemetery. These planview maps show subsurface structures at different depths. Sixty lines of data - individually representing vertical profiles - were collected and assembled as a 3-dimensional data array that can be horizontally "sliced" at different depths.)

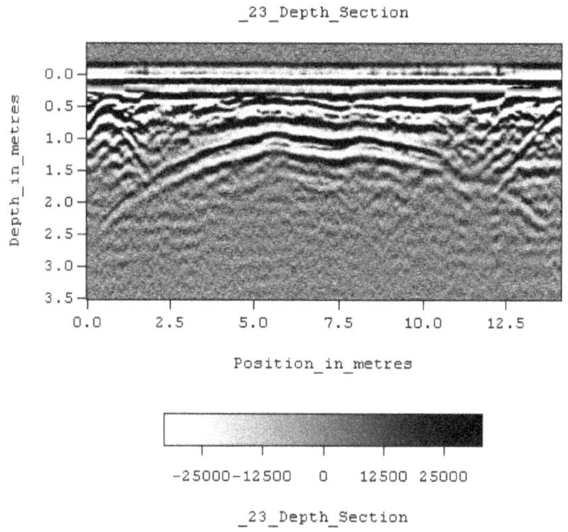

GPR depth section (profile) showing a single line of data from the survey of the historic crypt shown above. The domed roof of the crypt can be seen between 1 and 2.5 meters below surface.

GPR can be a powerful tool in favorable conditions (uniform sandy soils are ideal). Like other geophysical methods used in archaeology (and unlike excavation) it can locate artifacts and map features without any risk of damaging them. Among methods used in archaeological geophysics it is unique both in its ability to detect some small objects at relatively great depths, and in its ability to distinguish the depth of anomaly sources.

The principal disadvantage of GPR is that it is severely limited by less-than-ideal environmental conditions. Fine-grained sediments (clays and silts) are often problematic because their high electrical conductivity causes loss of signal strength; rocky or heterogeneous sediments scatter the GPR signal, weakening the useful signal while increasing extraneous noise.

In the field of cultural heritage GPR with high frequency antenna is also used for investigating historical masonry structures, detecting cracks and decay patterns of columns and detachment of frescoes.

Military

Military applications of ground-penetrating radar include detection of unexploded ordnance and detecting tunnels. In military applications and other common GPR applications, practitioners often use GPR in conjunction with other available geophysical techniques such as electrical resistivity and electromagnetic induction methods.

Vehicle Localization

A recent novel approach to vehicle localization using prior map based images from ground penetrating radar has been demonstrated. Termed "Localizing Ground Penetrating Radar" (LGPR), centimeter level accuracies at speeds up to 60 mph have been demonstrated.

Three-dimensional Imaging

Individual lines of GPR data represent a sectional (profile) view of the subsurface. Multiple lines of data systematically collected over an area may be used to construct three-dimensional or tomographic images. Data may be presented as three-dimensional blocks, or as horizontal or vertical slices. Horizontal slices (known as "depth slices" or "time slices") are essentially planview maps isolating specific depths. Time-slicing has become standard practice in archaeological applications, because horizontal patterning is often the most important indicator of cultural activities.

Limitations

The most significant performance limitation of GPR is in high-conductivity materials such as clay soils and soils that are salt contaminated. Performance is also limited by signal scattering in heterogeneous conditions (e.g. rocky soils).

Other disadvantages of currently available GPR systems include:

- Interpretation of radargrams is generally non-intuitive to the novice.

- Considerable expertise is necessary to effectively design, conduct, and interpret GPR surveys.

- Relatively high energy consumption can be problematic for extensive field surveys.

radar is sensitive to changes in material composition, detecting changes requires movement. When looking through stationary items using surface-penetrating or ground-penetrating radar, the equipment needs to be moved in order for the radar to examine the specified area by looking for differences in material composition. While it can identify items such as pipes, voids, and soil, it cannot identify the specific materials, such as gold and precious gems. It can however, be useful in providing subsurface mapping of potential gem-bearing pockets, or "vugs." The readings can be confused by moisture in the ground, and they can't separate gem-bearing pockets from the non-gem-bearing ones.

Antenna	Approximate Penetration in Dense Wet Clay	Approximate Penetration in Clean Dry Sand	Example of Smallest Visible Object
100 MHz	20 ft (6m)	60 ft+ (18m+)	Tunnel @ 60 ft (18m) depth 2 ft (60 cm) Pipe @ 20 ft (6m) depth
250 MHz	13 ft (4m)	40 ft (12m)	3 ft. (90 cm) Pipe @ 12m 6in. (15 cm) Pipe @ 13 ft (4m)
500 MHz	6 ft (1.8m)	14.5 ft. (4.4m)	4in. (10 cm) pipe @ 4m 3/16 in. (0.5 cm) hose 1.8m and less
1000 MHz	3 ft (90 cm)	6 ft (1.8m)	3/16 in. (0.5 cm) hose @ 3 ft. (90 cm) Wire mesh, shallow
2000 MHz	.5 ft (15 cm)	2 ft. (60 cm)	Monofilament fishing line

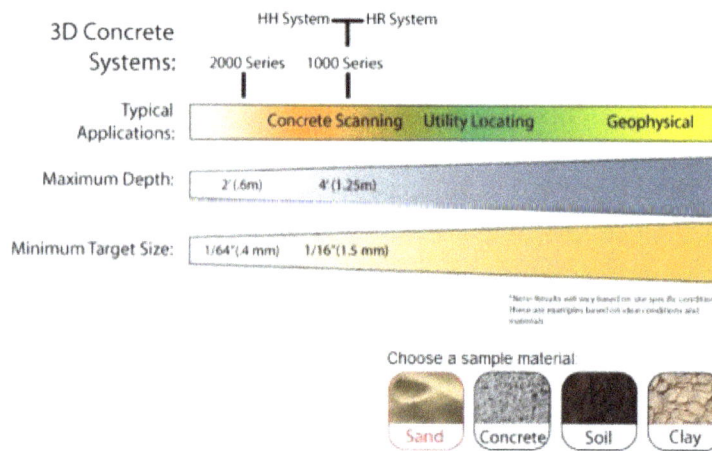

Ground penetrating radar examples.

When determining depth capabilities, the frequency range of the antenna dictates the size of the antenna and the depth capability. The grid spacing which is scanned is based on the size of the targets that need to be identified and the results required. Typical grid spacings can be 1 meter, 3 ft, 5 ft, 10 ft, 20 ft for ground surveys and 1in.-1 ft. for walls and floors.

The speed at which a radar signal travels is dependent upon the composition of the material being penetrated. The depth to a target is determined based on the amount of time it takes for the radar signal to reflect back to the unit's antenna. Radar signals travel at different velocities through different types of materials. It is possible to use the depth to a known object to determine a specific velocity and then calibrate the depth calculations.

Power Regulation

In 2005, the European Telecommunications Standards Institute introduced legislation to regulate GPR equipment and GPR operators to control excess emissions of electromagnetic radiation. The European GPR association (EuroGPR) was formed as a trade association to represent and protect the legitimate use of GPR in Europe.

Similar Technologies

Ground-penetrating radar uses a variety of technologies to generate the radar signal: these are impulse, stepped frequency, frequency-modulated continuous-wave (FMCW), and noise. Systems on the market in 2009 also use Digital signal processing (DSP) to process the data during survey work rather than off-line.

A special kind of GPR uses unmodulated continuous-wave signals. This holographic subsurface radar differs from other GPR types in that it records plan-view subsurface holograms. Depth penetration of this kind of radar is rather small (20–30 cm), but lateral resolution is enough to discriminate different types of landmines in the soil, or cavities, defects, bugging devices, or other hidden objects in walls, floors, and structural elements.

GPR is used on vehicles for close-in high-speed road survey and landmine detection as well as in stand-off mode.

Pipe-Penetrating Radar (PPR) is an application of GPR technologies applied in-pipe where the signals are directed through pipe and conduit walls to detect pipe wall thickness and voids behind the pipe walls.

Wall-penetrating radar can read through walls and even act as a motion sensor for police.

The "Mineseeker Project" seeks to design a system to determine whether landmines are present in areas using ultra wideband synthetic aperture radar units mounted on blimps.

Transient Electromagnetics

Transient electromagnetics, (also time-domain electromagnetics / TDEM), is a geophysical exploration technique in which electric and magnetic fields are induced by transient pulses of electric current and the subsequent decay response measured. TEM / TDEM methods are generally able to determine subsurface electrical properties, but are also sensitive to subsurface magnetic properties in applications like UXO detection and characterization. TEM/TDEM surveys area very common surface EM technique for mineral exploration, groundwater exploration, and for environmental mapping, used throughout the world in both onshore and offshore applications.

Physical Principles

Two fundamental electromagnetic principles are required to derive the physics behind TEM surveys: Faraday's law of induction and Lenz's Law. A loop of wire is generally energized by a direct current. At some time (t_o) the current is cut off as quickly as possible. Faraday's law dictates that a nearly identical current is induced in the subsurface to preserve the magnetic field produced by the original current (eddy currents). Due to ohmic losses, the induced surface currents dissipate—this causes a change in the magnetic field, which induces subsequent eddy currents. The net result is a downward and outward diffusion of currents in the subsurface which appear as an expanding smoke ring when the current density is contoured.

These currents produce a magnetic field by Faraday's law. At the surface, the change in magnetic field [flux] with time is measured. The way the currents diffuse in the subsurface is related to the conductivity distribution in the ground.

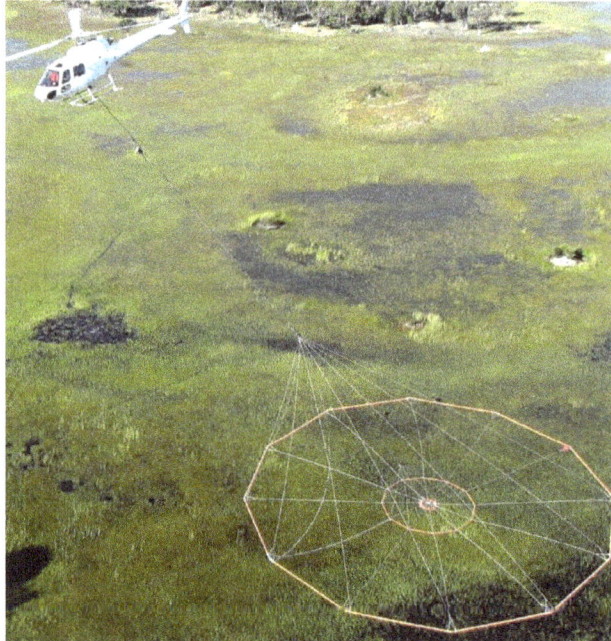

Helicopter Conducting TDEM Survey

This is a basic view of the physical principles involved. When conductive bodies are present, the diffusion of the transients is changed. In addition, transients are induced in the conductive bodies as well. This is only the most basic overview. The paper by McNeill is freely available from the Geonics website explaining the basics of the method.

TEM/TDEM Instrumentation and Sensors

TEM/TDEM systems consist of a transmitter instrument, transmitting coil or transmitting wire, receiver coil or antenna, and receiver instrument. Depending on subsurface resistivity, current induced, receiver sensitivity and transmitter-receiver geometry, TEM/TDEM measurements allow geophysical exploration from a few metres below the surface to several hundred metres of depth.

Low-power TEM/TDEM instruments can operate using C-cell batteries, and mid-range systems (approx. 2.5 kW) can operate with automotive batteries; more powerful systems (20 kW~150 kW) require truck-mounted generators to provide the necessary current for deep investigations.

Data Interpretation

For TEM/TDEM data interpretation the software realized 1D and 2D inversion are used (IX1d, Enigma, ZondTEM1d, TDEM Geomodel). Result of working same software - geoelectrical sections.

Commercial Applications

- Mining (mineral location and characterization)

- Groundwater characterization

- HVDC injection point mapping

- Oil and gas exploration

- Unexploded ordnance detection

References

- Nolet, G. (1987-01-01). Nolet, Guust, ed. Seismic wave propagation and seismic tomography. Seismology and Exploration Geophysics. Springer Netherlands. pp. 1–23. doi:10.1007/978-94-009-3899-1_1#page-1. ISBN 9789027725837.

- Daniels DJ (ed.) (2004). Ground Penetrating Radar (2nd ed.). Knoval (Institution of Engineering and Technology). pp. 1–4. ISBN 978-0-86341-360-5.

- "Seismic Tomography—Using earthquakes to image Earth's interior". Incorporated Research Institutions for Seismology (IRIS). Retrieved 18 May 2016.

- "Seismic Tomography" (PDF). earthscope.org. Incorporated Research Institutions for Seismology (IRIS). Retrieved 18 May 2016.

- Dziewonski, Adam. "Global Seismic Tomography: What we really can say and what we make up" (PDF). mantleplumes.org. Retrieved 18 May 2016.

- Cornick, Matthew; Koechling, Jeffrey; Stanley, Byron; Zhang, Beijia (2016-01-01). "Localizing Ground Penetrating RADAR: A Step Toward Robust Autonomous Ground Vehicle Localization". Journal of Field Robotics. 33 (1): 82–102. doi:10.1002/rob.21605. ISSN 1556-4967.

- Wilson, M. G. C.; Henry, G.; Marshall, T. R. (2006). "A review of the alluvial diamond industry and the gravels of the North West Province, South Africa" (PDF). South African Journal of Geology. Geological Society of South Africa. 109 (3): 301–314. doi:10.2113/gssajg.109.3.301. Retrieved 9 December 2012.

- "An analysis of magnetotelluric (MT) data over geothermal region of Bakreshwar, West Bengal". Cat.inist.fr. Retrieved 18 October 2011.

- "MT SURVEY IN TAIWAN EVALUATES THE POSSIBILITY OF CO2 SEQUESTRATION". Phoenix-geophysics.com. Retrieved 18 October 2011.

- "Characterizing a geothermal reservoir using broadband 2D MT survey in Theistareykir, Iceland" (PDF). Retrieved 18 October 2011.

- "Hydrogeologic assessment of the Amchitka Island nuclear test site (Alaska) with magnetotellurics". Cat.inist.fr. Retrieved 18 October 2011.

- "Welcome to the Scripps Institution of Oceanography Marine EM Laboratory website". Marineemlab.ucsd.edu. 23 April 2010. Retrieved 18 October 2011.

4

Well Logging: An Integrated Study

Well logging records formations under the earth that have been made in the process of digging a borehole. This chapter focuses on the various methods and techniques used in well logging like density logging, formation evaluation neutron porosity, sonic logging, gamma ray logging, spontaneous potential logging and mud logging. The chapter gives detailed information about the scope and service provided by each of the methods.

Well Logging

Well logging, also known as borehole logging is the practice of making a detailed record (a *well log*) of the geologic formations penetrated by a borehole. The log may be based either on visual inspection of samples brought to the surface (*geological* logs) or on physical measurements made by instruments lowered into the hole (*geophysical* logs). Some types of geophysical well logs can be done during any phase of a well's history: drilling, completing, producing, or abandoning. Well logging is performed in boreholes drilled for the oil and gas, groundwater, mineral and geothermal exploration, as well as part of environmental and geotechnical studies.

Wireline Logging

The oil and gas industry uses wireline logging to obtain a continuous record of a formation's rock properties. Wireline logging can be defined as being "The acquisition and analysis of geophysical data performed as a function of well bore depth, together with the provision of related services." Note that "wireline logging" and "mud logging" are not the same, yet are closely linked through the integration of the data sets. The measurements are made referenced to "TAH" - True Along Hole depth: these and the associated analysis can then be used to infer further properties, such as hydrocarbon saturation and formation pressure, and to make further drilling and production decisions.

Wireline logging is performed by lowering a 'logging tool' - or a string of one or more instruments - on the end of a wireline into an oil well (or borehole) and recording petrophysical properties using a variety of sensors. Logging tools developed over the years measure the natural gamma ray, electrical, acoustic, stimulated radioactive responses, electromagnetic, nuclear magnetic resonance, pressure and other properties of the rocks and their contained fluids. For this article, they are broadly broken down by the main property that they respond to.

The data itself is recorded either at surface (real-time mode), or in the hole (memory mode) to an electronic data format and then either a printed record or electronic presentation called a "well log" is provided to the client, along with an electronic copy of the raw data. Well logging operations can either be performed during the drilling process (Logging While Drilling), to provide real-time information about the formations being penetrated by the borehole, or once the well has reached Total Depth and the whole depth of the borehole can be logged.

Real-time data is recorded directly against measured cable depth. Memory data is recorded against time, and then depth data is simultaneously measured against time. The two data sets are then merged using the common time base to create an instrument response versus depth log. Memory recorded depth can also be corrected in exactly the same way as real-time corrections are made, so there should be no difference in the attainable TAH accuracy.

The measured cable depth can be derived from a number of different measurements, but is usually either recorded based on a calibrated wheel counter, or (more accurately) using magnetic marks which provide calibrated increments of cable length. The measurements made must then be corrected for elastic stretch and temperature.

There are many types of wireline logs and they can be categorized either by their function or by the technology that they use. "Open hole logs" are run before the oil or gas well is lined with pipe or cased. "Cased hole logs" are run after the well is lined with casing or production pipe.

Wireline logs can be divided into broad categories based on the physical properties measured.

History

Conrad and Marcel Schlumberger, who founded Schlumberger Limited in 1926, are considered the inventors of electric well logging. Conrad developed the Schlumberger array, which was a technique for prospecting for metal ore deposits, and the brothers adapted that surface technique to subsurface applications. On September 5, 1927, a crew working for Schlumberger lowered an electric sonde or tool down a well in Pechelbronn, Alsace, France creating the first well log. In modern terms, the first log was a resistivity log that could be described as 3.5-meter upside-down lateral log.

In 1931, Henri George Doll and G. Dechatre, working for Schlumberger, discovered that the galvanometer wiggled even when no current was being passed through the logging cables down in the well. This led to the discovery of the spontaneous potential (SP) which was as important as the ability to measure resistivity. The SP effect was produced naturally by the borehole mud at the boundaries of permeable beds. By simultaneously recording SP and resistivity, loggers could distinguish between permeable oil-bearing beds and impermeable nonproducing beds.

In 1940, Schlumberger invented the spontaneous potential dipmeter; this instrument allowed the calculation of the dip and direction of the dip of a layer. The basic dipmeter was later enhanced by the resistivity dipmeter (1947) and the continuous resistivity dipmeter (1952).

Oil-based mud (OBM) was first used in Rangely Field, Colorado in 1948. Normal electric logs require a conductive or water-based mud, but OBMs are nonconductive. The solution to this problem was the induction log, developed in the late 1940s.

The introduction of the transistor and integrated circuits in the 1960s made electric logs vastly more reliable. Computerization allowed much faster log processing, and dramatically expanded log data-gathering capacity. The 1970s brought more logs and computers. These included combo type logs where resistivity logs and porosity logs were recorded in one pass in the borehole.

The two types of porosity logs (acoustic logs and nuclear logs) date originally from the 1940s. Sonic logs grew out of technology developed during World War II. Nuclear logging has supplemented acoustic logging, but acoustic or sonic logs are still run on some combination logging tools.

Nuclear logging was initially developed to measure the natural gamma radiation emitted by underground formations. However, the industry quickly moved to logs that actively bombard rocks with nuclear particles. The gamma ray log, measuring the natural radioactivity, was introduced by Well Surveys Inc. in 1939, and the WSI neutron log came in 1941. The gamma ray log is particularly useful as shale beds which often provide a relatively low permeability cap over hydrocarbon reservoirs usually display a higher level of gamma radiation. These logs were important because they can be used in cased wells (wells with production casing). WSI quickly became part of Lane-Wells. During World War II, the US Government gave a near wartime monopoly on open-hole logging to Schlumberger, and a monopoly on cased-hole logging to Lane-Wells. Nuclear logs continued to evolve after the war.

The nuclear magnetic resonance log was developed in 1958 by Borg Warner. Initially, the NMR log was a scientific success but an engineering failure. However, the development of a continuous NMR logging tool by Numar (now a subsidiary of Halliburton) is a promising new technology.

Many modern oil and gas wells are drilled directionally. At first, loggers had to run their tools somehow attached to the drill pipe if the well was not vertical. Modern techniques now permit continuous information at the surface. This is known as logging while drilling (LWD) or measurement-while-drilling (MWD). MWD logs use mud pulse technology to transmit data from the tools on the bottom of the drillstring to the processors at the surface.

Electrical Logs

Resistivity Log

Resistivity logging measures the subsurface electrical resistivity, which is the ability to impede the flow of electric current. This helps to differentiate between formations filled with salty waters (good conductors of electricity) and those filled with hydrocarbons (poor conductors of electricity). Resistivity and porosity measurements are used to calculate water saturation. Resistivity is expressed in ohms or ohms\meter, and is frequently charted on a logarithm scale versus depth because of the large range of resistivity. The distance from the borehole penetrated by the current varies with the tool, from a few centimeters to one meter.

Borehole Imaging

The term "borehole imaging" refers to those logging and data-processing methods that are used to produce centimeter-scale images of the borehole wall and the rocks that make it up. The context is, therefore, that of open hole, but some of the tools are closely related to their cased-hole equivalents. Borehole imaging has been one of the most rapidly advancing technologies in wireline well logging. The applications range from detailed reservoir description through reservoir performance to enhanced hydrocarbon recovery. Specific applications are fracture identification, analysis of small-scale sedimentological features, evaluation of net pay in thinly bedded formations, and the identification of breakouts (irregularities in the borehole wall that are aligned with the minimum horizontal stress and appear where stresses around the wellbore exceed the compressive strength of the rock). The subject area can be classified into four parts:

1. Optical imaging

2. Acoustic imaging

3. Electrical imaging

4. Methods that draw on both acoustic and electrical imaging techniques using the same logging tool

Porosity Logs

Porosity logs measure the fraction or percentage of pore volume in a volume of rock. Most porosity logs use either acoustic or nuclear technology. Acoustic logs measure characteristics of sound waves propagated through the well-bore environment. Nuclear logs utilize nuclear reactions that take place in the downhole logging instrument or in the formation. Nuclear logs include density logs and neutron logs, as well as gamma ray logs which are used for correlation. The basic principle behind the use of nuclear technology is that a neutron source placed near the formation whose porosity is being measured will result in neutrons being scattered by the hydrogen atoms, largely those present in the formation fluid. Since there is little difference in the neutrons scattered by hydrocarbons or water, the porosity measured gives a figure close to the true physical porosity whereas the figure obtained from electrical resistivity measurements is that due to the conductive formation fluid. The difference between neutron porosity and electrical porosity measurements therefore indicates the presence of hydrocarbons in the formation fluid.

Density

The density log measures the bulk density of a formation by bombarding it with a radioactive source and measuring the resulting gamma ray count after the effects of Compton Scattering and Photoelectric absorption. This bulk density can then be used to determine porosity.

Neutron Porosity

The neutron porosity log works by bombarding a formation with high energy epithermal neutrons that lose energy through elastic scattering to near thermal levels before being absorbed by the nuclei of the formation atoms. Depending on the particular type of neutron logging tool, either the gamma ray of capture, scattered thermal neutrons or scattered, higher energy epithermal neutrons are detected. The neutron porosity log is predominantly sensitive to the quantity of hydrogen atoms in a particular formation, which generally corresponds to rock porosity.

Boron is known to cause anomalously low neutron tool count rates due to it having a high capture cross section for thermal neutron absorption. An increase in hydrogen concentration in clay minerals has a similar effect on the count rate.

Sonic

A sonic log provides a formation interval transit time, which typically varies lithology and rock texture but particularly porosity. The logging tool consists of a piezoelectric transmitter and receiver and the time taken to for the sound wave to travel the fixed distance between the two is recorded as an *interval transit time*.

Lithology Logs

Gamma Ray

A log of the natural radioactivity of the formation along the borehole, measured in API units, particularly useful for distinguishing between sands and shales in a siliclastic environment. This is because sandstones are usually nonradioactive quartz, whereas shales are naturally radioactive due to potassium isotopes in clays, and adsorbed uranium and thorium.

In some rocks, and in particular in carbonate rocks, the contribution from uranium can be large and erratic, and can cause the carbonate to be mistaken for a shale. In this case, The carbonate gamma ray is a better indicator of shaliness. the carbonate gamma ray log is a gamma ray log from which the uranium contribution has been subtracted.

Self/spontaneous Potential

The Spontaneous Potential (SP) log measures the natural or spontaneous potential difference between the borehole and the surface, without any applied current. It was one of the first wireline logs to be developed, found when a single potential electrode was lowered into a well and a potential was measured relative to a fixed reference electrode at the surface.

The most useful component of this potential difference is the electrochemical potential because it can cause a significant deflection in the SP response opposite permeable beds. The magnitude of this deflection depends mainly on the salinity contrast between the drilling mud and the formation water, and the clay content of the permeable bed. Therefore, the SP log is commonly used to detect permeable beds and to estimate clay content and formation water salinity.

Miscellaneous

Caliper

A tool that measures the diameter of the borehole, using either 2 or 4 arms. It can be used to detect regions where the borehole walls are compromised and the well logs may be less reliable.

Nuclear Magnetic Resonance

Nuclear magnetic resonance (NMR) logging uses the NMR response of a formation to directly determine its porosity and permeability, providing a continuous record along the length of the borehole. The chief application of the NMR tool is to determine moveable fluid volume (BVM) of a rock. This is the pore space excluding clay bound water (CBW) and irreducible water (BVI). Neither of these are moveable in the NMR sense, so these volumes are not easily observed on older logs. On modern tools, both CBW and BVI can often be seen in the signal response after transforming the relaxation curve to the porosity domain. Note that some of the moveable fluids (BVM) in the NMR sense are not actually moveable in the oilfield sense of the word. Residual oil and gas, heavy oil, and bitumen may appear moveable to the NMR precession measurement, but these will not necessarily flow into a well bore.

Spectral Noise Logging

Spectral noise logging (SNL) is an acoustic noise measuring technique used in oil and gas wells for well integrity analysis, identification of production and injection intervals and hydrodynamic characterisation of the reservoir. SNL records acoustic noise generated by fluid or gas flow through the reservoir or leaks in downhole well components.

Noise logging tools have been used in the petroleum industry for several decades. As far back as 1955, an acoustic detector was proposed for use in well integrity analysis to identify casing holes. Over many years, downhole noise logging tools proved effective in inflow and injectivity profiling of operating wells, leak detection, location of cross-flows behind casing, and even in determining reservoir fluid compositions. Robinson (1974) described how noise logging can be used to determine effective reservoir thickness.

Logging While Drilling

In the 1970s, a new approach to wireline logging was introduced in the form of logging while drilling (LWD). This technique provides similar well information to conventional wireline logging but instead of sensors being lowered into the well at the end of wireline cable, the sensors are integrated into the drill string and the measurements are made in real-time, whilst the well is being drilled. This allows drilling engineers and geologists to quickly obtain information such as porosity, resistivity, hole direction and weight-on-bit and they can use this information to make immediate decisions about the future of the well and the direction of drilling.

In LWD, measured data is transmitted to the surface in real time via pressure pulses in the well's mud fluid column. This mud telemetry method provides a bandwidth of less than 10 bits per second, although, as drilling through rock is a fairly slow process, data compression techniques mean that this is an ample bandwidth for real-time delivery of information. A higher sample rate of data is recorded into memory and retrieved when the drillstring is withdrawn at bit changes. High-definition downhole and subsurface information is available through networked or wired drillpipe that deliver memory quality data in real time.

Corrosion Well Logging

Throughout the life of the wells, integrity controles of the steel and cemented column (casing and tubing) are performed using calipers and thickness gauges. These advanced technical methods use non destructive technologies as ultrasonic, electromagnetic and magnetic transducers.

Memory Log

This method of data acquisition involves recording the sensor data into a down hole memory, rather than transmitting "Real Time" to surface. There are some advantages and disadvantages to this memory option.

- The tools can be conveyed into wells where the trajectory is deviated or extended beyond the reach of conventional Electric Wireline cables. This can involve a combination of weight to strength ratio of the electric cable over this extended reach. In such cases the memory tools can be conveyed on Pipe or Coil Tubing.

- The type of sensors are limited in comparison to those used on Electric Line, and tend to be focussed on the cased hole,production stage of the well. Although there are now developed some memory "Open Hole" compact formation evaluation tool combinations. These tools can be deployed and carried downhole concealed internally in drill pipe to protect them from damage while running in the hole, and then "Pumped" out the end at depth to initiate logging. Other basic open hole formation evaluation memory tools are available for use in "Commodity" markets on slickline to reduce costs and operating time.

- In cased hole operation there is normally a "Slick Line" intervention unit. This uses a solid mechanical wire (0.072 - 0.125 inches in OD), to manipulate or otherwise carry out operations in the well bore completion system. Memory operations are often carried out on this Slickline conveyance in preference to mobilizing a full service Electric Wireline unit.

- Since the results are not known until returned to surface, any realtime well dynamic changes cannot be monitored real time. This limits the ability to modify or change the well down hole production conditions accurately during the memory logging by changing the surface production rates. Something that is often done in Electric Line operations.

- Failure during recording is not known until the memory tools are retrieved. This loss of data can be a major issue on large offshore (expensive) locations. On land locations (e.g. South Texas, US) where there is what is called a "Commodity" Oil service sector, where logging often is without the rig infrastructure. this is less problematic, and logs are often run again without issue.

Coring

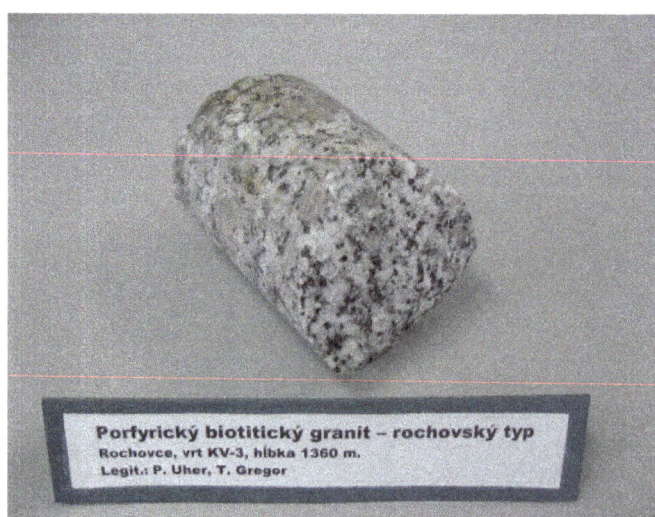

An example of a granite core

Coring is the process of obtaining an actual sample of a rock formation from the borehole. There are two main types of coring: 'full coring', in which a sample of rock is obtained using a specialised drill-bit as the borehole is first penetrating the formation and 'sidewall coring', in which multiple samples are obtained from the side of the borehole after it has penetrated through a formation. The main advantage of sidewall coring over full coring are that it is cheaper (drilling doesn't have to be stopped) and multiple samples can be easily acquired, with the main disadvantages being that there can be uncertainty in the depth at which the sample was acquired and the tool can fail to acquire the sample.

Mudlogging

Mud logs are well logs prepared by describing rock or soil cuttings brought to the surface by mud circulating in the borehole. In the oil industry they are usually prepared by a mud logging company contracted by the operating company. One parameter a typical mud log displays is the formation gas (gas units or ppm). "The gas recorder usually is scaled in terms of arbitrary gas units, which are defined differently by the various gas-detector manufactures. In practice, significance is placed only on relative changes in the gas concentrations detected." The current oil industry standard mud log normally includes real-time drilling parameters such as rate of penetration (ROP), lithology, gas hydrocarbons, flow line temperature (temperature of the drilling fluid) and chlorides but may also include mud weight, estimated pore pressure and corrected d-exponent (corrected drilling exponent) for a pressure pack log. Other information that is normally notated on a mud log include directional data (deviation surveys), weight on bit, rotary speed, pump pressure, pump rate, viscosity, drill bit info, casing shoe depths, formation tops, mud pump info, to name just a few.

Information Use

In the oil industry, the well and mud logs are usually transferred in 'real time' to the operating company, which uses these logs to make operational decisions about the well, to correlate formation depths with surrounding wells, and to make interpretations about the quantity and quality of hydrocarbons present. Specialists involved in well log interpretation are called log analysts.

Density Logging

Density logging is a well logging tool that can provide a continuous record of a formation's bulk density along the length of a borehole. In geology, bulk density is a function of the density of the minerals forming a rock (i.e. matrix) and the fluid enclosed in the pore spaces. This is one of three well logging tools that are commonly used to calculate porosity, the other two being sonic logging and neutron porosity logging

History & Principle

The tool was initially developed in the 1950s and was in use throughout the hydrocarbon industry by the 1960s. A type of active nuclear tool, a radioactive source and detector are lowered down the borehole and the source emits medium-energy gamma rays into the formation. Radioactive sources are typically a directional Cs-137 source. These gamma rays interact with electrons in the formation and are scattered in an interaction known as Compton scattering. The number of scattered gamma rays that reach the detector, placed at a set distance from the emitter, is related to the formation's electron density, which itself is related to the formation's bulk density (ρ_{bulk}) via

$$\rho_e = 2\rho_{bulk}\frac{Z}{A}$$

where Z is the atomic number, and A is the molecular weight of the compound. For most elements Z/A is about 1/2 (except for hydrogen where this ratio is 1). The electron density (ρ)$_e$ in g/cm³ determines the response of the density tool.

General Tool Design

The tool itself initially consisted of a radioactive source and a single detector, but this configuration is susceptible to the effects of the drilling fluid. In a similar way to how the sonic logging tool was improved to compensate for borehole effects, density logging now conventionally uses 2 or more detectors. In a 2 detector configuration, the short-spaced detector has a much shallower depth of investigation than the long-spaced detector so it is used to measure the effect that the drilling fluid has on the gamma ray detection. This result is then used to correct the long-spaced detector.

Formation Evaluation Neutron Porosity

In the field of formation evaluation, porosity is one of the key measurements to quantify oil and gas reserves. Neutron porosity measurement employs a neutron source to measure the hydrogen index in a reservoir, which is directly related to porosity. The Hydrogen Index (HI) of a material is defined as the ratio of the concentration of hydrogen atoms per cm³ in the material, to that of pure water at 75 °F. As hydrogen atoms are present in both water and oil filled reservoirs, measurement of the amount allows estimation of the amount of liquid-filled porosity.

Physics

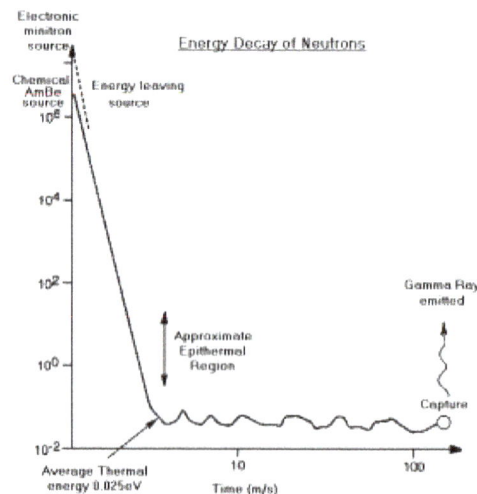

Fig1: Neutron Energy Decay

Neutrons are typically emitted by a radioactive source such as Americium Beryllium (Am-Be) or

Plutonium Beryllium (Pu-Be), or generated by electronic neutron generators such as minitron. Fast neutrons are emitted by these sources with energy ranges from 4 MeV to 14 MeV, and inelastically interact with matter. Once slowed down to 2 MeV, they start to scatter elastically and slow down further until the neutrons reach a thermal energy level of about 0.025 eV. When thermal neutrons are then absorbed, gamma rays are emitted. A suitable detector, positioned at a certain distance from the source, can measure either epithermal neutron population, thermal neutron population, or the gamma rays emitted after the absorption.

Mechanics of elastic collisions predict that the maximum energy transfer occurs during collisions of two particles of equal mass. Therefore, a hydrogen atom (H) will cause a neutron to slow down the most, as they are of roughly equal mass. As hydrogen is fundamentally associated to the amount of water and/or oil present in the pore space, measurement of neutron population within the investigated volume is directly linked to porosity.

Correction

Determination of porosity is one of the most important uses of neutron porosity log. Correction parameters for lithology, borehole parameters, and others are necessary for accurate porosity determination as follow:

1. Borehole size
2. Borehole salinity
3. Borehole temperature and pressure
4. Mud cake
5. Mud weight
6. Formation salinity
7. Tool standoff from borehole wall

Interpretation

Subject to various assumptions and corrections, values of apparent porosity can be derived from any neutron log. One can not underestimate the slow down of neutrons by other elements even if they are less effective. Certain effects, such as lithology, clay content, and amount and type of hydrocarbons, can be recognized and corrected for only if additional porosity information is available, for example from sonic and/or density log. Any interpretation of a neutron log alone should be undertaken with a realization of the uncertainties involved.

Effect of Light Hydrocarbon and Gas

The quantitative response of neutron tool to gas or light hydrocarbon depends primarily on hydrogen index and "excavation effect". The hydrogen index can be estimated from the composition and density of the hydrocarbons

Given a fixed volume, gas has considerably lower hydrogen concentration. When pore spaces in the rock are excavated and replaced with gas, the formation has smaller neutron-slowing charac-teristic, hence the terms "Excavation Effect". If this effect is ignored, a neutron log will show a low porosity value. This characteristic allows a neutron porosity log to be used with other porosity logs (such as a density log) to detect gas zones and identify gas-liquid contacts.

Measurement Technique

Neutron tools are based on the measurement of a neutron cloud of different energy levels within the investigated volume. Epithermal-neutron tools measure epithermal neutron density with en-ergy levels between 100eV and 0.1eV in the formation. Thermal-neutron tools only measure the population of neutrons with a thermal energy level, and Neutron-gamma tools measure the inten-sity of gamma flux generated by thermal neutron capture. The tools usually have two detectors (or more) with different spacings from the source to produce ratio of count rates, which theoretically reduce borehole effects.

A Helium-3 (He-3) filled proportional counter is the most common epithermal and thermal neu-tron detector. Helium has a high neutron capture cross section and produces the following reac-tion when interacting with a neutron.

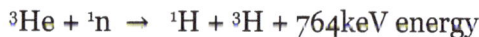

$$^3He + {}^1n \rightarrow {}^1H + {}^3H + 764keV \text{ energy}$$

To boost the charge produced by the interaction between Helium and a Neutron, a high voltage is applied to the anode of the counter. A high operating voltage is chosen to give enough gain for counting purposes. Most Helium-3 counters use a quench gas to stabilize high voltage perfor-mance and prevent run-away.

Sonic Logging

Sonic logging is a well logging tool that provides a formation's interval transit time, designated as Δt, which is a measure of a formation's capacity to transmit seismic waves. Geologically, this capacity varies with lithology and rock textures, most notably decreasing with an increasing effec-tive porosity. This means that a sonic log can be used to calculate the porosity of a formation if the seismic velocity of the rock matrix, V_{mat}, and pore fluid, V_l, are known, which is very useful for hydrocarbon exploration.

Process of Sonic Logging

The velocity is calculated by measuring the travel time from the piezoelectric transmitter to the receiver, normally with the units microsecond per foot (a measure of slowness). To compensate for the variations in the drilling mud thickness, there are actually two receivers, one near and one far. This is because the travel time within the drilling mud will be common for both, so the travel time within the formation is given by:

$$\Delta t = t_{far} - t_{near} \, ;$$

where t_{far} = travel time to far receiver; t_{near} = travel time to near receiver.

If it is necessary to compensate for tool tilt and variations in the borehole width then both up-down and down-up arrays can be used and an average can be calculated. Overall this gives a sonic log that can be made up of 1 or 2 pulse generators and 2 or 4 detectors, all located in single unit called a "sonde", which is lowered down the well.

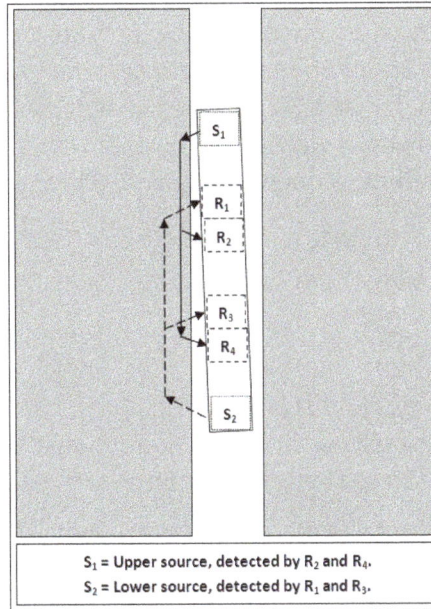

S_1 = Upper source, detected by R_2 and R_4.
S_2 = Lower source, detected by R_1 and R_3.

Source and receiver relationships for a sonic log

An additional way in which the sonic log tool can be altered is increasing or decreasing the separation between the source and receivers. This gives deeper penetration and overcomes the problem of low velocity zones posed by borehole wall damage.

Cycle Skipping

The returning signal is a wavetrain and not a sharp pulse, so the detectors are only activated at a certain signal threshold. Sometimes, both detectors won't be activated by the same peak (or trough) and the next peak (or trough) wave will activate one of them instead. This type of error is called cycle skipping and is easily identified because the time difference is equal to the time interval between successive pulse cycles.

Calculating Porosity

Many relationships between travel time and porosity have been proposed, the most commonly accepted is the Wyllie time-average equation. The equation basically holds that the total travel time recorded on the log is the sum of the time the sonic wave spends travelling the solid part of the rock, called the rock matrix and the time spent travelling through the fluids in the pores. This equation is empirical and makes no allowance for the structure of the rock matrix or the connectivity of the pore spaces so extra corrections can often be added to it. The Wyllie time-average equation is:

$$\frac{1}{V} = \frac{\phi}{V_f} + \frac{1-\phi}{V_{mat}}$$

where V = seismic velocity of the formation; V_f = seismic velocity of the pore fluid; V_{mat} = seismic velocity of the rock matrix; ϕ = porosity.

Accuracy

The accuracy of sonic logs is rather poor, evident by the fact that regular- and long-spaced log measurements often conflict, and this should be taken into account when there are disagreements between seismic data and sonic log data. Other considerations include that the resolution of sonic logs is on a scale of inches whereas seismic reflection data resolution is on the scale of meters, and the two methods use significantly different frequency ranges so travel times will vary due to dispersion.

In order to investigate how the varying size of a borehole has affected a sonic log, the results can be plotted against those of a caliper log.

Calibrated Sonic Log

To improve the tie between well data and seismic data a "check-shot" survey is often used to generate a calibrated sonic log. A geophone, or array of geophones is lowered down the borehole, with a seismic source located at the surface. The seismic source is fired with the geophone(s) at a series of different depths, with the interval transit times being recorded. This is often done during the acquisition of a vertical seismic profile.

Use in Mineral Exploration

Sonic logs are also used in mineral exploration, especially exploration for iron and potassium.

Gamma Ray Logging

Gamma ray logging is a method of measuring naturally occurring gamma radiation to characterize the rock or sediment in a borehole or drill hole. It is a wireline logging method used in mining, mineral exploration, water-well drilling, for formation evaluation in oil and gas well drilling and for other related purposes. Different types of rock emit different amounts and different spectra of natural gamma radiation. In particular, shales usually emit more gamma rays than other sedimentary rocks, such as sandstone, gypsum, salt, coal, dolomite, or limestone because radioactive potassium is a common component in their clay content, and because the cation exchange capacity of clay causes them to absorb uranium and thorium. This difference in radioactivity between shales and sandstones/carbonate rocks allows the gamma tool to distinguish between shales and non-shales.

The gamma ray log, like other types of well logging, is done by lowering an instrument down the drill hole and recording gamma radiation variation with depth. In the United States, the device most commonly records measurements at 1/2-foot intervals. Gamma radiation is usually recorded in API units, a measurement originated by the petroleum industry. Gamma logs are attenuated by diameter of the borehole because of the properties of the fluid filling the borehole, but because gamma logs are most often used in a qualitative way, corrections are usually not necessary.

Well logging methods
• Spontaneous potential
• **Gamma ray**
• Resistivity
• Density
• Sonic
• Caliper
• Mud
• While drilling (LWD)
• Measurement (MWD)
• Nuclear magnetic resonance (NMR)

Three elements and their decay chains are responsible for the radiation emitted by rock: potassium, thorium and uranium. Shales often contain potassium as part of their clay content, and tend to absorb uranium and thorium as well. A common gamma-ray log records the total radiation and cannot distinguish between the radioactive elements, while a spectral gamma ray log can.

For standard GR logs, the value measured is calculated from thorium in ppm, Uranium in ppm and potassium in percent. GR API = 8 × Uranium concentration in ppm + 4 × thorium concentration in ppm + 16 × potassium concentration in percent. Due to the weight of uranium concentration in the calculation, anomalous concentrations of uranium can cause clean sand reservoirs to appear shaley. Spectral Gamma ray is used to provide an individual reading for each element so anomalies in concentration can be found and interpreted.

An advantage of the gamma log over some other types of well logs is that it works through the steel and cement walls of cased boreholes. Although concrete and steel absorb some of the gamma radiation, enough travels through the steel and cement to allow qualitative determinations.

Sometimes non-shales also have elevated levels of gamma radiation. Sandstone can contain uranium mineralization, potassium feldspar, clay filling, or rock fragments that cause it to have higher-than usual gamma readings. Coal and dolomite may contain absorbed uranium. Evaporite deposits may contain potassium minerals such as carnallite. When this is the case, spectral gamma ray logging can be done to identify these anomalies.

Spectral Logging

The technique of measuring the spectrum, or number and energy, of gamma rays emitted as natural radioactivity by the formation. There are three sources of natural radioactivity in the Earth: 40K, 232Th and 238U, or potassium, thorium and uranium. These radioactive isotopes emit gamma rays that have characteristic energy levels. The quantity and energy of these gamma rays can be measured in a scintillation detector. A log of natural gamma ray spectroscopy is usually presented as a total gamma ray log and the weight fraction of potassium (%), thorium (ppm) and uranium (ppm). The primary standards for the weight fractions are formations with known quantities of the three isotopes. Natural gamma ray spectroscopy logs were introduced in the early 1970s, although they had been studied from the 1950s.

The characteristic gamma ray line that is associated with each component:

- Potassium : Gamma ray energy 1.46 MeV

- Thorium series: Gamma ray energy 2.61 MeV

- Uranium-Radium series: Gamma ray energy 1.76 MeV

Another example of the use of spectral gamma ray logs is to identify specific clay types, like Kaolinite or Illite. This can be used for environmental interpretation as Kaolinite forms from Feldspars in tropic soils by leaching of Potassium; and low Potassium readings may thus indicate paleosols. The identification of clay types is also useful for calculating the effective porosity of reservoir rock.

Use in Mineral Exploration

Gamma ray logs are also used in mineral exploration, especially exploration for phosphates, uranium, and potassium salts.

Spontaneous Potential Logging

Spontaneous Potential Log (SP)

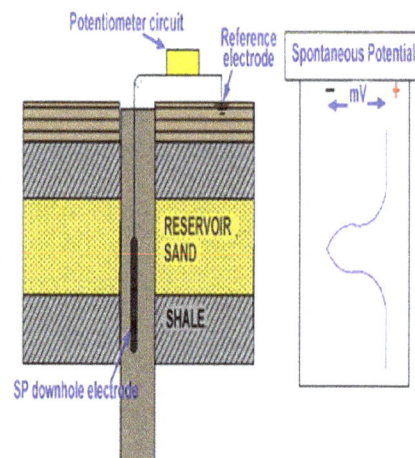

- The spontaneous potential (SP) curve records the naturally occurring electrical potential (voltage) produced by the interaction of formation connate water, conductive drilling fluid, and shale
- The SP curve reflects a difference in the electrical potential between a movable electrode in the borehole and a fixed reference electrode at the surface
- Though the SP is used primarily as a lithology indicator and as a correlation tool, it has other uses as well:
  - permeability indicator,
  - shale volume indicator
  - porosity indicator, and
  - measurement of Rw (hence formation water salinity).

The spontaneous potential log, commonly called the self potential log or SP log, is a passive measurement taken by oil industry well loggers to characterise rock formation properties. The log works by measuring small electric potentials (measured in millivolts) between depths in the borehole and a grounded electrode at the surface. Conductive bore hole fluids are necessary to create a SP response, so the SP log cannot be used in nonconductive drilling muds (e.g. oil-based mud) or air filled holes.

The change in voltage through the well bore is caused by a buildup of charge on the well bore walls. Clays and shales (which are composed predominantly of clays) will generate one charge and permeable formations such as sandstone will generate an opposite one. Spontaneous potentials occur when two aqueous solutions with different ionic concentrations are placed in contact through a porous, semi-permeable membrane. In nature, ions tend to migrate from high to low ionic concentrations. In the case of SP logging, the two aqueous solutions are the well bore fluid (drilling mud) and the formation water (connate water). The potential opposite shales is called the baseline, and typically shifts only slowly over the depth of the borehole.

The relative salinity of the mud and the formation water will determine the which way the SP curve will deflect opposite a permeable formation. Generally if the ionic concentration of the well bore fluid is less than the formation fluid then the SP reading will be more negative (usually plotted as a deflection to the left). If the formation fluid has an ionic concentration less than the well bore fluid, the voltage deflection will be positive (usually plotted as an excursion to the right). The amplitudes of the line made by the changing SP will vary from formation to formation and will not give a definitive answer to how permeable or the porosity of the formation that it is logging.

The presence of hydrocarbons (e.g. oil, natural gas, condensate) will reduce the response on an SP log because the interstitial water contact with the well bore fluid is reduced. This phenomena is called hydrocarbon suppression and can be used to diagnose rocks for commercial potential. The SP curve is usually 'flat' opposite shale formations because there is no ion exchange due to the low permeability, low porosity properties (tight)thus creating a baseline. Tight rocks other than shale (e.g. tight sandstones, tight carbonates) will also result in poor or no response on the SP curve because of no ion exchange.

The SP tool is one of the simplest tools and is generally run as standard when logging a hole, along with the gamma ray. SP data can be used to find:

- Depths of permeable formations

- The boundaries of these formations

- Correlation of formations when compared with data from other analogue wells

- Values for the formation-water resistivity

The SP curve can be influenced by various factors both in the formation and introduced into the wellbore by the drilling process. These factors can cause the SP curve to be muted or even inverted depending on the situation.

- Formation bed thickness

- Resistivities in the formation bed and the adjacent formations

- Resistivity and make up of the drilling mud

- Wellbore diameter

- The depth of invasion by the drilling mud into the formation

Mud invasion into the permeable formation can cause the deflections in the SP curve to be rounded off and to reduce the amplitude of thin beds.

A smaller wellbore will cause, like a mud filtrate invasion, the deflections on the SP curve to be rounded off and decrease the amplitude opposite thin beds, while a larger diameter wellbore has the opposite effect. If the salinity of the mud filtrate is greater than formation water the SP currents will flow in opposite direction.In that case SP deflection will be positive towards to the right. Positive deflections are observed for fresh water bearing formations.

Mud Logging

Well logging methods
• Spontaneous potential
• Gamma ray
• Resistivity
• Density
• Sonic
• Caliper
• **Mud**
• While drilling (LWD)
• Measurement (MWD)
• Nuclear magnetic resonance (NMR)

Mud logging is the creation of a detailed record (well log) of a borehole by examining the cuttings of rock brought to the surface by the circulating drilling medium (most commonly drilling mud). Mud logging is usually performed by a third-party mud logging company. This provides well owners and producers with information about the lithology and fluid content of the borehole while drilling. Historically it is the earliest type of well log. Under some circumstances compressed air is employed as a circulating fluid, rather than mud. Although most commonly used in petroleum exploration, mud logging is also sometimes used when drilling water wells and in other mineral exploration, where drilling fluid is the circulating medium used to lift cuttings out of the hole. In hydrocarbon exploration, hydrocarbon surface gas detectors record the level of natural gas brought up in the mud. A mobile laboratory is situated by the mud logging company near the drilling rig or on deck of an offshore drilling rig, or on a drill ship.

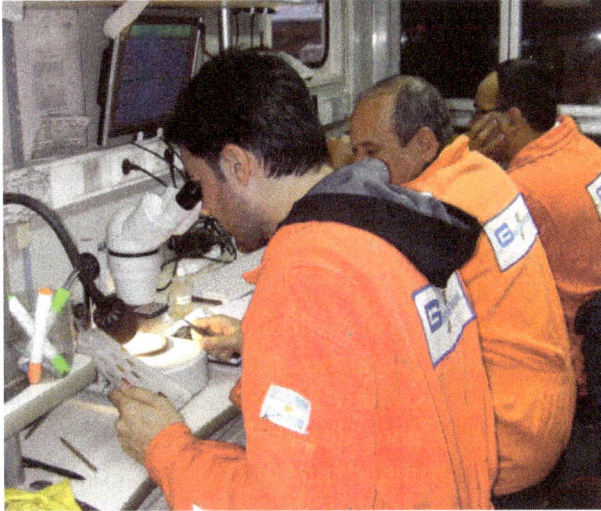

Inside a Mud Logging cabin

The Service Provided

Mud logging technicians in an oil field drilling operation determine positions of hydrocarbons with respect to depth, identify downhole lithology, monitor natural gas entering the drilling mud stream, and draw well logs for use by oil company geologist. Rock cuttings circulated to the surface in drilling mud are sampled and analyzed.

The mud logging company is normally contracted by the oil company (or operator). They then organize this information in the form of a graphic log, showing the data charted on a representation of the wellbore.

Well-site geologist mudlogging

The oil company representative (Company Man or "CoMan") together with the tool pusher, and well-site geologist (WSG) provides mud loggers their instruction. The mud logging company is contracted specifically as to when to start well-logging activity and what services to provide. Mud logging may begin on the first day of drilling, known as the "spud in" date but is more likely at some later time (and depth) determined by the oil industry geologist's research. The mud logger may also possess logs from wells drilled in the surrounding area. This information (known as "offset data") can provide valuable clues as to the characteristics of the particular geo-strata that the rig crew is about to drill through.

Mud loggers connect various sensors to the drilling apparatus and install specialized equipment to monitor or "log" drill activity. This can be physically and mentally challenging, especially when having to be done during drilling activity. Much of the equipment will require precise calibration or alignment by the mud logger to provide accurate readings.

Mud logging technicians observe and interpret the indicators in the mud returns during the drilling process, and at regular intervals log properties such as drilling rate, mud weight, flowline temperature, oil indicators, pump pressure, pump rate, lithology (rock type) of the drilled cuttings, and other data. Mud logging requires a good deal of diligence and attention. Sampling the drilled cuttings must be performed at predetermined intervals, and can be difficult during rapid drilling.

Another important task of the mud logger is to monitor gas levels (and types) and notify other personnel on the rig when gas levels may be reaching dangerous levels, so appropriate steps can be taken to avoid a dangerous well blowout condition. Because of the lag time between drilling and the time required for the mud and cuttings to return to the surface, a modern augmentation has come into use: Measurement while drilling. The MWD technician, often a separate service company employee, logs data in a similar manner but the data is different in source and content. Most of the data logged by an MWD technician comes from expensive and complex, sometimes electronic, tools that are downhole installed at or near the drill bit.

Scope

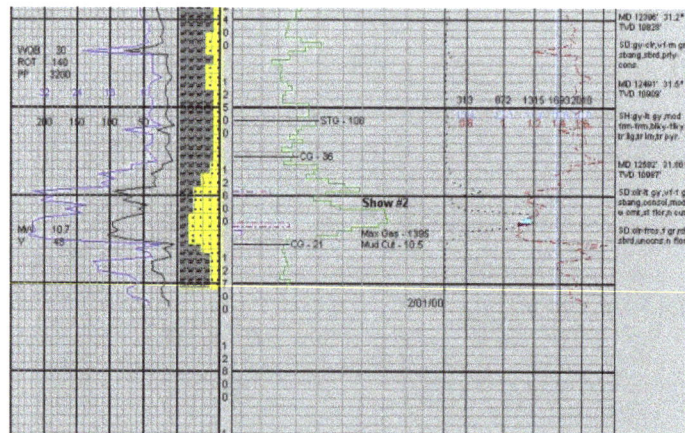

1" (5 foot average) mud log showing heavy (hydrocarbons) (large area of yellow)

Mud logging includes observation and microscopic examination of drill cuttings (formation rock chips), and evaluation of gas hydrocarbon and its constituents, basic chemical and mechanical parameters of drilling fluid or drilling mud (such as chlorides and temperature), as well as compiling other information about the drilling parameters. Then data is plotted on a graphic log called a mud log. Example1, Example2.

Other real-time drilling parameters that may be compiled include, but are not limited to; rate of penetration (ROP) of the bit (sometimes called the drill rate), pump rate (quantity of fluid being pumped), pump pressure, weight on bit, drill string weight, rotary speed, rotary torque, RPM (Revolutions Per Minute), SPM (Strokes Per Minute) mud volumes, mud weight and mud viscosity. This information is usually obtained by attaching *monitoring devices* to the drilling rig's equipment with a few exceptions such as the mud weight and mud viscosity which are measured by the derrickhand or the mud engineer.

Rate of drilling is affected by the pressure of the column of mud in the borehole and its relative counterbalance to the internal pore pressures of the encountered rock. A rock pressure greater than the mud fluid will tend to cause rock fragments to spall as it is cut and can increase the drilling rate. "D-exponents" are mathematical trend lines which estimate this internal pressure. Thus both visual evidence of spalling and mathematical plotting assist in formulating recommendations for optimum drilling mud densities for both safety (blowout prevention) and economics. (Faster drilling is generally preferred.)

1" (every foot) mud log showing corrected d-Exponent trending into pressure above the sand

Mud logging is often written as a single word "mudlogging". The finished product can be called a "mud log" or "mudlog". The occupational description is "mud logger" or "mudlogger". In most cases, the two word usage seems to be more common. The mud log provides a reliable *time log* of drilled formations.

Details of the Mud Log

- The *rate of penetration* in (Figure 1 & 2) is represented by the black line on the left side of the log. The farther to the left that the line goes, the faster the rate of penetration. On this mud log, ROP is measured in feet per hour but on some older, hand drawn mud logs, it is measured in minutes per foot.

- The *porosity* in (Figure 1) is represented by the blue line farthest to the left of the log. It indicates the pore space within the rock structure. An analogy would be the holes in a sponge.

The oil and gas resides within this pore space. Notice how far to the left the porosity goes where all the sand (in yellow) is. This indicates that the sand has good porosity. Porosity is not a direct or physical measurement of the pore space but rather an extrapolation from other drilling parameters and therefore not always reliable.

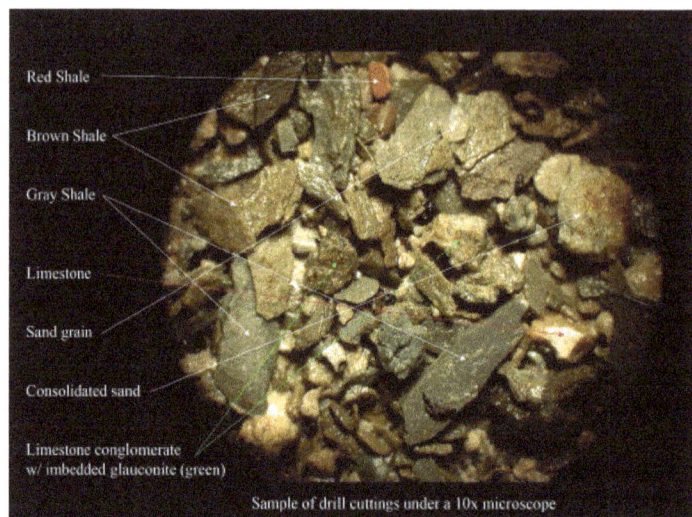

(Figure 3) Sample of drill cuttings of shale while drilling an oil well in Louisiana. For reference, the sand grain and red shale are approximately 2 mm. in dia.

- The *lithology* in (Figure 1 & 2) is represented by the cyan, gray/black and yellow blocks of color. Cyan = lime, gray/black = shale and yellow = sand. More yellow represents more sand identified at that depth. The lithology is measured as percentage of the total sample, as visually inspected under a microscope, normally at 10x magnification (Figure 3). These are but a fraction of the different types of formations that might be encountered. (Color coding is not necessarily standardized among different mud logging companies, though the symbol representation for each are very similar.) In (Figure 3) you can see a sample of cuttings under a microscope at 10x magnification after they have been washed off. Some of the larger shale and lime fragments are separated from this sample by running it through sieves and must be considered when estimating percentages. Also, this image view is only a fragment of the total sample and some of the sand at the bottom of the tray can not be seen and must also be considered in the total estimation. With that in mind this sample would be considered to be about 90% shale, 5% sand and 5% lime (In 5% increments).

- The *gas* in (Figure 1 & 2) is represented by the green line and is measured in units of ppm (parts per million) as the quantity of total gas, but does not represent the actual quantity of oil or gas the reservoir contains. In (Figure 1) the squared-off dash-dot lines just to the right of the sand (in yellow) and left of the gas (in green) represents the heavier hydrocarbons detected. Cyan = C_2 (ethane), purple = C_3 (propane) and blue = C_4 (butane). Detecting and analyzing these heavy gases help to determine the type of oil or gas the formation contains.

References

- Etnyre, L.M. (1989). Finding Oil and Gas from Well Logs. Kluwer Academic Publishers. p. 249 p. ISBN 978-0442223090.

- Bourgoyne, Adam; Keith Millheim, Martin Chenevert, F.S. Young Jr. (1986). Applied Drilling Engineering. Richardson, TX: Society of Petroleum Engineers. p. 274 p. ISBN 1-55563-001-4.

- Sengel, E.W. "Bill" (1981). Handbook on well logging. Oklahoma City, Oklahoma: Institute for Energy Development. p. 168 p. ISBN 0-89419-112-8.

- Ali, T.H.; M. Sas; J.H. Hood; S.R. Lemke; A. Srinivasan (2008). "High Speed Telemetry Drill Pipe Network Optimizes Drilling Dynamics And Wellbore Placement". Society of Petroleum Engineers. Retrieved 25 September 2012.

5

A Comprehensive Study of Oil Well and its Techniques

An oil well is usually drilled to access deposits of oil and gas that are available below the surface of the Earth. Sometimes there can be mishaps in the drilling and extraction of hydrocarbons from the oil well. In order to avoid such contingencies from occurring, measures have to be taken. This chapter explores the facets of oil well control, hydraulic fracturing and offshore drilling and their fundamental terminology. There is a section on oil platform as well.

Oil Well

An oil well is a boring in the Earth that is designed to bring petroleum oil hydrocarbons to the surface. Usually some natural gas is produced along with the oil. A well that is designed to produce mainly or only gas may be termed a gas well.

The pumpjack, such as this one located south of Midland, Texas, is a common sight in West Texas

History

The earliest known oil wells were drilled in China in 347 CE. These wells had depths of up to about 240 metres (790 ft) and were drilled using bits attached to bamboo poles. The oil was burned to evaporate brine and produce salt. By the 10th century, extensive bamboo pipelines connected oil wells with salt springs. The ancient records of China and Japan are said to contain many allusions to the use of natural gas for lighting and heating. Petroleum was known as *Burning water* in Japan in the 7th century.

According to Kasem Ajram, petroleum was distilled by the Persian alchemist Muhammad ibn Za-karīya Rāzi (Rhazes) in the 9th century, producing chemicals such as kerosene in the alembic (*al-ambiq*), and which was mainly used for kerosene lamps. Arab and Persian chemists also distilled crude oil in order to produce flammable products for military purposes. Through Islamic Spain, distillation became available in Western Europe by the 12th century.

Bottom Part of an Oil Drilling Derrick in Brazoria County, Texas (Harry Walker Photograph, circa 1940)

Some sources claim that from the 9th century, oil fields were exploited in the area around modern Baku, Azerbaijan, to produce naphtha for the petroleum industry. These fields were described by Marco Polo in the 13th century, who described the output of those oil wells as hundreds of shiploads. When Marco Polo in 1264 visited the Azerbaijani city of Baku, on the shores of the Caspian Sea, he saw oil being collected from seeps. He wrote that "on the confines toward Geirgine there is a fountain from which oil springs in great abundance, in as much as a hundred shiploads might be taken from it at one time."

A FOUNTAIN AT BIBI-EIBAT IN FLAMES, BAKU

1904 oil well fire at Bibi-Eibat

In North America, the first commercial oil well entered operation in Oil Springs, Ontario in 1858, while the first offshore oil well was drilled in 1896 at the Summerland Oil Field on the California Coast.

The earliest oil wells in modern times were drilled percussively, by repeatedly raising and dropping a cable tool into the earth. In the 20th century, cable tools were largely replaced with rotary drilling, which could drill boreholes to much greater depths and in less time. The record-depth Kola Borehole used non-rotary mud motor drilling to achieve a depth of over 12,000 metres (39,000 ft).

Until the 1970s, most oil wells were vertical, although lithological and mechanical imperfections cause most wells to deviate at least slightly from true vertical. However, modern directional drilling technologies allow for strongly deviated wells which can, given sufficient depth and with the proper tools, actually become horizontal. This is of great value as the reservoir rocks which contain hydrocarbons are usually horizontal or nearly horizontal; a horizontal wellbore placed in a production zone has more surface area in the production zone than a vertical well, resulting in a higher production rate. The use of deviated and horizontal drilling has also made it possible to reach reservoirs several kilometers or miles away from the drilling location (extended reach drilling), allowing for the production of hydrocarbons located below locations that are either difficult to place a drilling rig on, environmentally sensitive, or populated.

Life of a Well

A schematic of a typical oil well being produced by a pumpjack, which is used to produce the remaining recoverable oil after natural pressure is no longer sufficient to raise oil to the surface

The creation and life of a well can be divided up into five segments:

- Planning
- Drilling
- Completion

- Production

- Abandonment

Drilling

The well is created by drilling a hole 12 cm to 1 meter (5 in to 40 in) in diameter into the earth with a drilling rig that rotates a drill string with a bit attached. After the hole is drilled, sections of steel pipe (casing), slightly smaller in diameter than the borehole, are placed in the hole. Cement may be placed between the outside of the casing and the borehole known as the annulus. The casing provides structural integrity to the newly drilled wellbore, in addition to isolating potentially dangerous high pressure zones from each other and from the surface.

With these zones safely isolated and the formation protected by the casing, the well can be drilled deeper (into potentially more-unstable and violent formations) with a smaller bit, and also cased with a smaller size casing. Modern wells often have two to five sets of subsequently smaller hole sizes drilled inside one another, each cemented with casing.

To Drill the well

Well Casing

- The drill bit, aided by the weight of thick walled pipes called "drill collars" above it, cuts into the rock. There are different types of drill bit; some cause the rock to disintegrate by compressive failure, while others shear slices off the rock as the bit turns.

- Drilling fluid, a.k.a. "mud", is pumped down the inside of the drill pipe and exits at the drill bit. The principal components of drilling fluid are usually water and clay, but it also typically contains a complex mixture of fluids, solids and chemicals that must be carefully tailored to provide the correct physical and chemical characteristics required to safely drill the well. Particular functions of the drilling mud include cooling the bit, lifting rock cuttings to the surface, preventing destabilisation of the rock in the wellbore walls and overcoming the pressure of fluids inside the rock so that these fluids do not enter the wellbore. Some oil wells are drilled with air or foam as the drilling fluid.

Mud log in process, a common way to study the lithology when drilling oil wells

- The generated rock "cuttings" are swept up by the drilling fluid as it circulates back to surface outside the drill pipe. The fluid then goes through "shakers" which strain the cuttings from the good fluid which is returned to the pit. Watching for abnormalities in the returning cuttings and monitoring pit volume or rate of returning fluid are imperative to catch "kicks" early. A "kick" is when the formation pressure at the depth of the bit is more than the hydrostatic head of the mud above, which if not controlled temporarily by closing the blowout preventers and ultimately by increasing the density of the drilling fluid would allow formation fluids and mud to come up through the annulus uncontrollably.

- The pipe or drill string to which the bit is attached is gradually lengthened as the well gets deeper by screwing in additional 9 m (30 ft) sections or "joints" of pipe under the kelly or topdrive at the surface. This process is called making a connection, or "tripping". Joints can be combined for more efficient tripping when pulling out of the hole by creating stands of multiple joints. A conventional triple, for example, would pull pipe out of the hole three joints at a time and stack them in the derrick. Many modern rigs, called "super singles", trip pipe one at a time, laying it out on racks as they go.

This process is all facilitated by a drilling rig which contains all necessary equipment to circulate the drilling fluid, hoist and turn the pipe, control downhole, remove cuttings from the drilling fluid, and generate on-site power for these operations.

Completion

After drilling and casing the well, it must be 'completed'. Completion is the process in which the well is enabled to produce oil or gas.

In a cased-hole completion, small holes called perforations are made in the portion of the casing which passed through the production zone, to provide a path for the oil to flow from the surrounding rock into the production tubing. In open hole completion, often 'sand screens' or a 'gravel pack' is installed in the last drilled, uncased reservoir section. These maintain structural integrity of the wellbore in the absence of casing, while still allowing flow from the reservoir into the wellbore. Screens also control the migration of formation sands into production tubulars and surface equipment, which can cause washouts and other problems, particularly from unconsolidated sand formations of offshore fields.

Modern drilling rig in Argentina

After a flow path is made, acids and fracturing fluids may be pumped into the well to fracture, clean, or otherwise prepare and stimulate the reservoir rock to optimally produce hydrocarbons into the wellbore. Finally, the area above the reservoir section of the well is packed off inside the casing, and connected to the surface via a smaller diameter pipe called tubing. This arrangement provides a redundant barrier to leaks of hydrocarbons as well as allowing damaged sections to be replaced. Also, the smaller cross-sectional area of the tubing produces reservoir fluids at an increased velocity in order to minimize liquid fallback that would create additional back pressure, and shields the casing from corrosive well fluids.

In many wells, the natural pressure of the subsurface reservoir is high enough for the oil or gas to flow to the surface. However, this is not always the case, especially in depleted fields where the pressures have been lowered by other producing wells, or in low permeability oil reservoirs. Installing a smaller diameter tubing may be enough to help the production, but artificial lift methods may also be needed. Common solutions include downhole pumps, gas lift, or surface pump jacks. Many new systems in the last ten years have been introduced for well completion. Multiple packer systems with frac ports or port collars in an all in one system have cut completion costs and improved production, especially in the case of horizontal wells. These new systems allow casings to run into the lateral zone with proper packer/frac port placement for optimal hydrocarbon recovery.

Production

The production stage is the most important stage of a well's life; when the oil and gas are produced. By this time, the oil rigs and workover rigs used to drill and complete the well have moved off the wellbore, and the top is usually outfitted with a collection of valves called a Christmas tree or production tree. These valves regulate pressures, control flows, and allow access to the wellbore in case further completion work is needed. From the outlet valve of the production tree, the flow can be connected to a distribution network of pipelines and tanks to supply the product to refineries, natural gas compressor stations, or oil export terminals.

As long as the pressure in the reservoir remains high enough, the production tree is all that is required to produce the well. If the pressure depletes and it is considered economically viable, an artificial lift method mentioned in the completions section can be employed.

Workovers are often necessary in older wells, which may need smaller diameter tubing, scale or paraffin removal, acid matrix jobs, or completing new zones of interest in a shallower reservoir. Such remedial work can be performed using workover rigs – also known as *pulling units*, *completion rigs* or "service rigs" – to pull and replace tubing, or by the use of well intervention techniques utilizing coiled tubing. Depending on the type of lift system and wellhead a rod rig or flushby can be used to change a pump without pulling the tubing.

Enhanced recovery methods such as water flooding, steam flooding, or CO_2 flooding may be used to increase reservoir pressure and provide a "sweep" effect to push hydrocarbons out of the reservoir. Such methods require the use of injection wells (often chosen from old production wells in a carefully determined pattern), and are used when facing problems with reservoir pressure depletion, high oil viscosity, or can even be employed early in a field's life. In certain cases – depending on the reservoir's geomechanics – reservoir engineers may determine that ultimate recoverable oil may be increased by applying a waterflooding strategy early in the field's development rather than later. Such enhanced recovery techniques are often called "tertiary recovery".

Types of Well

A natural gas well in the southeast Lost Hills Field, California, US.

Raising the derrick

Oil extraction in Boryslav in 1909

Fossil-fuel wells come in many varieties. By produced fluid, there can be wells that produce oil, wells that produce oil *and* natural gas, or wells that *only* produce natural gas. Natural gas is almost always a byproduct of producing oil, since the small, light gas carbon chains come out of solution as they undergo pressure reduction from the reservoir to the surface, similar to uncapping a bottle of soda pop where the carbon dioxide effervesces. Unwanted natural gas can be a disposal problem at the well site. If there is not a market for natural gas near the wellhead it is virtually valueless since it must be piped to the end user. Until recently, such unwanted gas was burned off at the wellsite, but due to environmental concerns this practice is becoming less common. Often, unwanted (or 'stranded' gas without a market) gas is pumped back into the reservoir with an 'injection' well for disposal or repressurizing the producing formation. Another solution is to export the natural gas as a liquid. Gas to liquid, (GTL) is a developing technology that converts stranded natural gas into synthetic gasoline, diesel or jet fuel through the Fischer-Tropsch process developed in World War II Germany. Such fuels can be transported through conventional pipelines and tankers to users. Proponents claim GTL fuels burn cleaner than comparable petroleum fuels. Most major international oil companies are in advanced development stages of GTL production, e.g. the 140,000 bbl/d (22,000 m³/d) Pearl GTL plant in Qatar, scheduled to come online in 2011. In locations such as the United States with a high natural gas demand, pipelines are constructed to take the gas from the wellsite to the end consumer.

Another obvious way to classify oil wells is by land or offshore wells. There is very little difference in the well itself. An offshore well targets a reservoir that happens to be underneath an ocean. Due to logistics, drilling an offshore well is far more costly than an onshore well. By far the most common type is the onshore well. These wells dot the Southern and Central Great Plains, Southwestern United States, and are the most common wells in the Middle East.

Another way to classify oil wells is by their purpose in contributing to the development of a resource. They can be characterized as:

- *wildcat wells* are drilled where little or no known geological information is available. The site may have been selected because of wells drilled some distance from the proposed location but on a terrain that appeared similar to the proposed site.

- *exploration wells* are drilled purely for exploratory (information gathering) purposes in a new area, the site selection is usually based on seismic data, satellite surveys etc. Details gathered in this well includes the presence of Hydrocarbon in the drilled location, the amount of fluid present and the depth at which oil or/and gas occurs.

- *appraisal wells* are used to assess characteristics (such as flow rate, reserve quantity) of a proven hydrocarbon accumulation. The purpose of this well is to reduce uncertainty about the characteristics and properties of the hydrocarbon present in the field.

- *production wells* are drilled primarily for producing oil or gas, once the producing structure and characteristics are determined.

- *development wells* are wells drilled for the production of oil or gas already proven by appraisal drilling to be suitable for exploitation.

- *Abandoned well* are wells permanently plugged in the drilling phase for technical reasons.

At a producing well site, active wells may be further categorised as:

- *oil producers* producing predominantly liquid hydrocarbons, but mostly with some associated gas.

- *gas producers* producing almost entirely gaseous hydrocarbons.

- *water injectors* injecting water into the formation to maintain reservoir pressure, or simply to dispose of water produced with the hydrocarbons because even after treatment, it would be too oily and too saline to be considered clean for dumping overboard offshore, let alone into a fresh water resource in the case of onshore wells. Water injection into the producing zone frequently has an element of reservoir management; however, often produced water disposal is into shallower zones safely beneath any fresh water zones.

- *aquifer producers* intentionally producing water for re-injection to manage pressure. If possible this water will come from the reservoir itself. Using aquifer produced water rather than water from other sources is to preclude chemical incompatibility that might lead to reservoir-plugging precipitates. These wells will generally be needed only if produced water from the oil or gas producers is insufficient for reservoir management purposes.

- *gas injectors* injecting gas into the reservoir often as a means of disposal or sequestering for later production, but also to maintain reservoir pressure.

Lahee Classification

- *New Field Wildcat* (NFW) – far from other producing fields and on a structure that has not previously produced.

- *New Pool Wildcat* (NPW) – new pools on already producing structure.

- *Deeper Pool Test* (DPT) – on already producing structure and pool, but on a deeper pay zone.

- *Shallower Pool Test* (SPT) – on already producing structure and pool, but on a shallower pay zone.

- *Outpost* (OUT) – usually two or more locations from nearest productive area.
- *Development Well* (DEV) – can be on the extension of a pay zone, or between existing wells (*Infill*).

Cost

The cost of a well depends mainly on the daily rate of the drilling rig, the extra services required to drill the well, the duration of the well program (including downtime and weather time), and the remoteness of the location (logistic supply costs).

The daily rates of offshore drilling rigs vary by their capability, and the market availability. Rig rates reported by industry web service show that the deepwater water floating drilling rigs are over twice that of the shallow water fleet, and rates for jackup fleet can vary by factor of 3 depending upon capability.

With deepwater drilling rig rates in 2015 of around $520,000/day, and similar additional spread costs, a deep water well of duration of 100 days can cost around US$100 million.

With high performance jackup rig rates in 2015 of around $177,000, and similar service costs, a high pressure, high temperature well of duration 100 days can cost about US$30 million.

Onshore wells can be considerably cheaper, particularly if the field is at a shallow depth, where costs range from less than $1 million to $15 million for deep and difficult wells.

The total cost of an oil well mentioned does not include the costs associated with the risk of explosion and leakage of oil. Those costs include the cost of protecting against such disasters, the cost of the cleanup effort, and the hard-to-calculate cost of damage to the company's image.

Reefs

Offshore platforms (the structure supporting the wells) often provide habitat for marine life. After the wells have been abandoned, sometimes the platforms can be toppled in place or moved elsewhere to be dropped to the ocean floor to produce artificial reefs.

Oil Well Control

Oil well control is the management of the dangerous effects caused by the unexpected release of formation fluid, such as natural gas and/or crude oil, upon surface equipment of oil or gas drilling rigs and escaping into the atmosphere. Technically, oil well control involves preventing the formation fluid, usually referred to as kick, from entering into the wellbore during drilling.

Formation fluid can enter the wellbore if the pressure exerted by the column of drilling fluid is not great enough to overcome the pressure exerted by the fluids in the formation being drilled. Oil well control also includes monitoring a well for signs of impending influx of formation fluid into the wellbore during drilling and procedures, to stop the well from flowing when it happens by taking proper remedial actions.

Failure to manage and control these pressure effects can cause serious equipment damage and injury, or loss of life. Improperly managed well control situations can cause blowouts, which are uncontrolled and explosive expulsions of formation fluid from the well, potentially resulting in a fire.

Importance of Oil Well Control

Modern driller Argentina.

Oil well control is one of the most important aspects of drilling operations. Improper handling of kicks in oil well control can result in blowouts with very grave consequences, including the loss of valuable resources. Even though the cost of a blowout (as a result of improper/no oil well control) can easily reach several millions of US dollars, the monetary loss is not as serious as the other damages that can occur: irreparable damage to the environment, waste of valuable resources, ruined equipment, and most importantly, the safety and lives of personnel on the drilling rig.

In order to avert the consequences of blowout, the utmost attention must be given to oil well control. That is why oil well control procedures should be in place prior to the start of an abnormal situation noticed within the wellbore, and ideally when a new rig position is sited. In other words, this includes the time the new location is picked, all drilling, completion, workover, snubbing and any other drilling-related operations that should be executed with proper oil well control in mind. This type of preparation involves widespread training of personnel, the development of strict operational guidelines and the design of drilling programs — maximizing the probability of successfully regaining hydrostatic control of a well after a significant influx of formation fluid has taken place. dangers of oil drilling

Fundamental Concepts and Terminology

Pressure is a very important concept in the oil and gas industry. Pressure can be defined as: the force exerted per unit area. Its SI unit is newtons per square metre or pascals. Another unit, bar, is also widely used as a measure of pressure, with 1 bar equal to 100 kilopascals. Normally pressure is measured in the U.S. petroleum industry in units of pounds force per square inch of area, or psi. 1000 psi equals 6894.76 kilo-pascals.

Hydrostatic Pressure

Hydrostatic pressure (HSP), as stated, is defined as pressure due to a column of fluid that is not moving. That is, a column of fluid that is static, or at rest, exerts pressure due to local force of gravity on the column of the fluid.

The formula for calculating hydrostatic pressure in SI units (N/m^2) is:

Hydrostatic pressure = Height (m) × Density (kg/m^3) × Gravity (m/s^2).

All fluids in a wellbore exert hydrostatic pressure, which is a function of density and vertical height of the fluid column. In US oil field units, hydrostatic pressure can be expressed as:

HSP = 0.052 × *MW* × *TVD*', where MW (Mud Weight or density) is the drilling-fluid density in pounds per gallon (ppg), TVD is the true vertical depth in feet and HSP is the hydrostatic pressure in psi.

The 0.052 is needed as the conversion factor to psi unit of HSP.

To convert these units to SI units, one can use:

- 1 ppg = ≈ 119.8264273 kg/m^3
- 1 ft = 0.3048 metres
- 1 psi = 0.0689475729 bar
- 1 bar = 10^5 pascals

Pressure Gradient

The pressure gradient is described as the pressure per unit length. Often in oil well control, pressure exerted by fluid is expressed in terms of its pressure gradient. The SI unit is pascals/metre. The hydrostatic pressure gradient can be written as:

Pressure gradient (psi/ft) = HSP/TVD = 0.052 × MW (ppg).

Formation Pressure

Formation pressure is the pressure exerted by the formation fluids, which are the liquids and gases contained in the geologic formations encountered while drilling for oil or gas. It can also be said to be the pressure contained within the pores of the formation or reservoir being drilled. Formation pressure is a result of the hydrostatic pressure of the formation fluids, above the depth of interest, together with pressure trapped in the formation. Under formation pressure, there are 3 levels: normally pressured formation, abnormal formation pressure, or subnormal formation pressure.

Normally Pressured Formation

Normally pressured formation has a formation pressure that is the same with the hydrostatic pressure of the fluids above it. As the fluids above the formation are usually some form of water, this pressure can be defined as the pressure exerted by a column of water from the formation's depth to sea level.

The normal hydrostatic pressure gradient for freshwater is 0.433 pounds per square inch per foot (psi/ft), or 9.792 kilopascals per meter (kPa/m), and 0.465 psi/ft for water with dissolved solids like in Gulf Coast waters, or 10.516 kPa/m. The density of formation water in saline or marine environments, such as along the Gulf Coast, is about 9.0 ppg or 1078.43 kg/m³. Since this is the highest for both Gulf Coast water and fresh water, a normally pressured formation can be controlled with a 9.0 ppg mud.

Sometimes the weight of the overburden, which refers to the rocks and fluids above the formation, will tend to compact the formation, resulting in pressure built-up within the formation if the fluids are trapped in place. The formation in this case will retain its normal pressure only if there is a communication with the surface. Otherwise, an *abnormal formation pressure* will result.

Abnormal Formation Pressure

As discussed above, once the fluids are trapped within the formation and not allow to escape there is a pressure build-up leading to abnormally high formation pressures. This will generally require a mud weight of greater than 9.0 ppg to control. Excess pressure, called "overpressure" or "geopressure", can cause a well to blow out or become uncontrollable during drilling.

Subnormal Formation Pressure

Subnormal formation pressure is a formation pressure that is less than the normal pressure for the given depth. It is common in formations that had undergone production of original hydrocarbon or formation fluid in them.

Overburden Pressure

Overburden pressure is the pressure exerted by the weight of the rocks and contained fluids above the zone of interest. Overburden pressure varies in different regions and formations. It is the force that tends to compact a formation vertically. The density of these usual ranges of rocks is about 18 to 22 ppg (2,157 to 2,636 kg/m³). This range of densities will generate an overburden pressure gradient of about 1 psi/ft (22.7 kPa/m). Usually, the 1 psi/ft is not applicable for shallow marine sediments or massive salt. In offshore however, there is a lighter column of sea water, and the column of underwater rock does not go all the way to the surface. Therefore, a lower overburden pressure is usually generated at an offshore depth, than would be found at the same depth on land.

Mathematically, overburden pressure can be derived as:

$$S = \rho_b \times D \times g$$

where

   g = acceleration due to gravity

   S = overburden pressure

   ρ_b = average formation bulk density

   D = vertical thickness of the overlying sediments

The bulk density of the sediment is a function of rock matrix density, porosity within the confines of the pore spaces, and porefluid density. This can be expressed as

$$\rho_b = \varphi\rho_f + (1 - \varphi)\rho_m$$

where

φ = rock porosity

ρ_f = formation fluid density

ρ_m = rock matrix density

Fracture Pressure

Fracture pressure can be defined as pressure required to cause a formation to fail or split. As the name implies, it is the pressure that causes the formation to fracture and the circulating fluid to be lost. Fracture pressure is usually expressed as a gradient, with the common units being psi/ft (kPa/m) or ppg (kg/m³).

To fracture a formation, three things are generally needed, which are:

1. Pump into the formation. This will require a pressure in the wellbore greater than formation pressure.

2. The pressure in the wellbore must also exceed the rock matrix strength.

3. And finally the wellbore pressure must be greater than one of the three principal stresses in the formation.

Pump Pressure (System Pressure Losses)

Pump pressure, which is also referred to as *system pressure loss*, is the sum total of all the pressure losses from the oil well surface equipment, the drill pipe, the drill collar, the drill bit, and annular friction losses around the drill collar and drill pipe. It measures the system pressure loss at the start of the circulating system and measures the total friction pressure.

Slow Pump Pressure (SPP)

Slow pump pressure is the circulating pressure (pressure used to pump fluid through the whole active fluid system, including the borehole and all the surface tanks that constitute the primary system during drilling) at a reduced rate. SPP is very important during a well kill operation in which circulation (a process in which drilling fluid is circulated out of the suction pit, down the drill pipe and drill collars, out the bit, up the annulus, and back to the pits while drilling proceeds) is done at a reduced rate to allow better control of circulating pressures and to enable the mud properties (density and viscosity) to be kept at desired values. The slow pump pressure can also be referred to as "kill rate pressure" or "slow circulating pressure" or "kill speed pressure" and so on.

Shut-in Drill Pipe Pressure

Shut-in drill pipe pressure (SIDPP), which is recorded when a well is shut in on a kick, is a measure of the difference between the pressure at the bottom of the hole and the hydrostatic pressure (HSP) in the drillpipe. During a well shut-in, the pressure of the wellbore stabilizes, and the formation pressure equals the pressure at the bottom of the hole. The drillpipe at this time should be full of known-density fluid. Therefore, the formation pressure can be easily calculated using the SIDPP. This means that the SIDPP gives a direct of formation pressure during a kick.

Shut-in Casing Pressure (SICP)

The *shut-in casing pressure* (SICP) is a measure of the difference between the formation pressure and the HSP in the annulus when a kick occurs.

The pressures encountered in the annulus can be estimated using the following mathematical equation:

$$FP = HSP_{mud} + HSP_{influx} + SICP$$

where

FP = formation pressure (psi)

HSP_{mud} = Hydrostatic pressure of the mud in the annulus (psi)

HSP_{influx} = Hydrostatic pressure of the influx (psi)

SICP = shut-in casing pressure (psi)

Bottom-hole Pressure (BHP)

Bottom-hole pressure (BHP) is the pressure at the bottom of a well. The pressure is usually measured at the bottom of the hole. This pressure may be calculated in a static, fluid-filled wellbore with the equation:

$$BHP = D \times \rho \times C,$$

where

BHP = bottom-hole pressure

D = the vertical depth of the well

ρ = density

C = units conversion factor

(or, in the English system, BHP = D × MWD × 0.052).

In Canada the formula is depth in meters x density in kgs x the constant gravity factor (0.00981), which will give the hydrostatic pressure of the well bore or (hp) hp=bhp with pumps off. The bottom-hole pressure is dependent on the following:

- Hydrostatic pressure (HSP)
- Shut-in surface pressure (SIP)

- Friction pressure

- Surge pressure (occurs when transient pressure increases the bottom-hole pressure)

- Swab pressure (occurs when transient pressure reduces the bottom-hole pressure)

Therefore, BHP can be said to be the sum of all pressures at the bottom of the wellhole, which equals:

BHP = HSP + SIP + friction + Surge - swab

Basic Calculations in Oil Well Control

There are some basic calculations that need to be carried during oil well control. A few of these essential calculations will be discussed below. Most of the units here are in US oil field units, but these units can be converted to their SI units equivalent by using this Conversion of units link.

Capacity

The capacity of drill string is an essential issue in oil well control. The capacity of drillpipe, drill collars or hole is the volume of fluid that can be contained within them.

The capacity formula is as shown below:

Capacity = $ID^2/1029.4$

where

Capacity = Volume in barrels per foot(bbl/ft)

ID = Inside diameter in inches

1029.4 = Units conversion factor

Also the total pipe or hole volume is given by :

Volume in barrels (bbls) = *Capacity* (bbl/ft) × length (ft)

Feet of pipe occupied by a given volume is given by:

Feet of pipe (ft) = *Volume of mud* (bbls) / *Capacity* (bbls/ft)

Capacity calculation is important in oil well control due to the following:

- Volume of the drillpipe and the drill collars must be pumped to get kill weight mud to the bit during kill operation.

- It is used to spot pills and plugs at various depths in the wellbore.

Annular Capacity

This is the volume contained between the inside diameter of the hole and the outside diameter of the pipe. Annular capacity is given by :

$$Annular\ capacity\ (bbl/ft) = (ID_{hole}^2 - OD_{pipe}^2) / 1029.4$$

where

ID_{hole}^2 = Inside diameter of the casing or open hole in inches

OD_{pipe}^2 = Outside diameter of the pipe in inches

Similarly

$$Annular\ volume\ (bbls) = Annular\ capacity\ (bbl/ft) \times length\ (ft)$$

and

Feet occupied by volume of mud in annulus = Volume of mud (bbls) / *Annular Capacity* (bbls/ft).

Fluid Level Drop

Fluid level drop is the distance the mud level will drop when a dry string(a bit that is not plugged) is being pulled from the wellbore and it is given by:

$$Fluid\ level\ drop = Bbl\ disp / (CSG\ cap - Pipe\ disp)$$

or

$$Fluid\ level\ drop = Bbl\ disp / (Ann\ cap + Pipe\ cap)$$

and the resulting loss of HSP is given by:

$$Lost\ HSP = 0.052 \times MW \times Fluid\ drop$$

where

Fluid drop = distance the fluid falls (ft)

Bbl disp = displacement of the pulled pipe (bbl)

CSG cap = casing capacity (bbl/ft)

Pipe disp = pipe displacement (bbl/ft)

Ann cap = Annular capacity between casing and pipe (bbl/ft)

Pipe cap = pipe capacity

Lost HSP = Lost hydrostatic pressure (psi)

MW = mud weight (ppg)

When pulling a wet string (the bit is plugged) and the fluid from the drillpipe is not returned to the hole. The fluid drop is then changed to the following:

$$Fluid\ level\ drop = Bbl\ disp / Ann\ cap$$

Kill Weight Fluid

Kill weight fluid which can also be called *Kill weight Mud* is the density of the mud required to balance formation pressure during kill operation. The Kill Weight Mud can be calculated by:

$$KWM = SIDPP/(0.052 \times TVD) + OWM$$

where

KWM = kill weight mud (ppg)

$SIDPP$ = shut-in drillpipe pressure (psi)

TVD = true vertical depth (ft)

OWM = original weight mud (ppg)

But when the formation pressure can be determined from data sources such as bottom hole pressure, then *KWM* can be calculated as follows:

$$KWM = FP / 0.052 \times TVD$$

where *FP* = Formation pressure.

Kicks

Ixtoc I oil well blowout

Kick is the entry of formation fluid into the wellbore during drilling operations. It occurs because the pressure exerted by the column of drilling fluid is not great enough to overcome the pressure exerted by the fluids in the formation drilled. The whole essence of oil well control is to prevent kick from occurring and if it happens to prevent it from developing into blowout. An uncontrolled kick usually results from not deploying the proper equipment, using poor practices, or a lack of training of the rig crews. Loss of oil well control may lead into blowout, which represents one of the most severe threats associated with the exploration of petroleum resources involving the risk of lives and environmental and economic consequences.

Causes of Kicks

A kick will occur when the bottom hole pressure(BHP) of a well falls below the formation pressure and the formation fluid flows into the wellbore.There are usually causes for kicks some of which are:

- Failure to keep the hole full during a trip

- Swabbing while tripping

- Lost circulation

- Insufficient density of fluid

- Abnormal pressure

- Drilling into an adjacent well

- Lost control during drill stem test

Failure to Keep the Hole Full During a Trip

Tripping is the complete operation of removing the drillstring from the wellbore and running it back in the hole. This operation is typically undertaken when the bit (which is the tool used to crush or cut rock during drilling) becomes dull or broken, and no longer drills the rock efficiently. A typical drilling operation of deep oil or gas wells may require up to 8 or more trips of the drill string to replace a dull rotary bit for one well.

Tripping out of the hole means that the entire volume of steel (of drillstring) is being removed, or has been removed, from the well. This displacement of the drill string (the steel) will leave out a volume of space that must be replaced with an equal volume of mud. If the replacement is not done, the fluid level in the wellbore will drop, resulting in a loss of hydrostatic pressure (HSP) and bottom hole pressure (BHP). If this bottom hole pressure reduction goes below the formation pressure, a kick will definitely occur.

Swabbing While Tripping

Swabbing occurs when bottom hole pressure is reduced due to the effects of pulling the drill string upward in the bored hole. During the tripping out of the hole, the space formed by the drillpipe, drill collar, or tubing (which are being removed) must be replaced by something, usually mud. If the rate of tripping out is greater than the rate the mud is being pumped into the void space (created by the removal of the drill string), then swab will occur. If the reduction in bottom hole pressure caused by swabbing is below formation pressure, then a kick will occur.

Lost Circulation

Lost circulation usually occurs when the hydrostatic pressure fractures an open formation. When this occurs, there is loss in circulation, and the height of the fluid column decreases, leading to lower HSP in the wellbore. A kick can occur if steps are not taken to keep the hole full. Lost circulation can be caused by:

- excessive mud weights

- excessive annular friction loss

- excessive surge pressure during trips, or "spudding" the bit

- excessive shut-in pressures.

Insufficient Density of Fluid

If the density of the drilling fluid or mud in the well bore is not sufficient to keep the formation pressure in check, then a kick can occur. Insufficient density of the drilling fluid can be as a result of the following :

- attempting to drill by using an underbalanced weight solution

- excessive dilution of the mud

- heavy rains in the pits

- barite settling in the pits

- spotting low density pills in the well.

Abnormal Pressure

Another cause of kicks is drilling accidentally into abnormally-pressured permeable zones. The increased formation pressure may be greater than the bottom hole pressure, resulting in a kick.

Drilling into an Adjacent Well

Drilling into an adjacent well is a potential problem, particularly in offshore drilling where a large number of directional wells are drilled from the same platform. If the drilling well penetrates the production string of a previously completed well, the formation fluid from the completed well will enter the wellbore of the drilling well, causing a kick. If this occurs at a shallow depth, it is an extremely dangerous situation and could easily result in an uncontrolled blowout with little to no warning of the event.

Lost Control During Drill Stem Test

A drill-stem test is performed by setting a packer above the formation to be tested, and allowing the formation to flow. During the course of the test, the bore hole or casing below the packer, and at least a portion of the drill pipe or tubing, is filled with formation fluid. At the conclusion of the test, this fluid must be removed by proper well control techniques to return the well to a safe condition. Failure to follow the correct procedures to kill the well could lead to a blowout.

Kick Warning Signs

In oil well control, a kick should be able to be detected promptly, and if a kick is detected, proper kick prevention operations must be taken immediately to avoid a blowout. There are various telltale signs that signal an alert crew that a kick is about to start. Knowing these signs will keep a kicking oil well under control, and avoid a blowout:

Deepwater Horizon drilling rig blowout, 21 April 2010

Sudden Increase in Drilling Rate

A sudden increase in penetration rate (drilling break) is usually caused by a change in the type of formation being drilled. However, it may also signal an increase in formation pore pressure, which may indicate a possible kick.

Increase in Annulus Flow Rate

If the rate at which the pumps are running is held constant, then the flow from the annulus should be constant. If the annulus flow increases without a corresponding change in pumping rate, the additional flow is caused by formation fluid(s) feeding into the well bore or gas expansion. This will indicate an impending kick.

Gain in Pit Volume

If there is an unexplained increase in the volume of surface mud in the pit (a large tank that holds drilling fluid on the rig), it could signify an impending kick. This is because as the formation fluid feeds into the wellbore, it causes more drilling fluid to flow from the annulus than is pumped down the drill string, thus the volume of fluid in the pit(s) increases.

Change in Pump Speed/pressure

A decrease in pump pressure or increase in pump speed can happen as a result of a decrease in hydrostatic pressure of the annulus as the formation fluids enters the wellbore. As the lighter formation fluid flows into the wellbore, the hydrostatic pressure exerted by the annular column of fluid decreases, and the drilling fluid in the drill pipe tends to U-tube into the annulus. When this occurs, the pump pressure will drop, and the pump speed will increase. The lower pump pressure and increase in pump speed symptoms can also be indicative of a hole in the drill string, commonly referred to as a washout. Until a confirmation can be made whether a washout or a well kick has occurred, a kick should be assumed.

Improper Fill on Trips

Improper fill on trip occurs when the volume of drilling fluid to keep the hole full on a Trip (complete operation of removing the drillstring from the wellbore and running it back in the hole) is less than that calculated or less than Trip Book Record. This condition is usually caused by formation fluid entering the wellbore due to the swabbing action of the drill string, and, if action is not taken soon, the well will enter a kick state.

Categories of Oil Well Control

There are basically three types of oil well control which are: primary oil well control, secondary oil well control, and tertiary oil well control. Those types are explained below.

Primary Oil Well Control

Primary oil well control is the process which maintains a hydrostatic pressure in the wellbore greater than the pressure of the fluids in the formation being drilled, but less than formation fracture pressure. It uses the mud weight to provide sufficient pressure to prevent an influx of formation fluid into the wellbore. If hydrostatic pressure is less than formation pressure, then formation fluids will enter the wellbore. If the hydrostatic pressure of the fluid in the wellbore exceeds the fracture pressure of the formation, then the fluid in the well could be lost. In an extreme case of lost circulation, the formation pressure may exceed hydrostatic pressure, allowing formation fluids to enter into the well.

Secondary Oil Well Control

Secondary oil well control is done after the Primary oil well control has failed to prevent formation fluids entering the wellbore. This process is stopped using a "blow out preventer", a BOP, to prevent the escape of wellbore fluids from the well. As the rams and choke of the BOP remain closed, a pressure built up test is carried out and a kill mud weight calculated and pumped inside the well to kill the kick and circulate it out.

Tertiary (or Shearing) Oil Well Control

Tertiary oil well control describes the third line of defense, where the formation cannot be controlled by primary or secondary well control (hydrostatic and equipment). This happens in underground blowout situations. The following are examples of tertiary well control:

- Drill a relief well to hit an adjacent well that is flowing and kill the well with heavy mud

- Rapid pumping of heavy mud to control the well with equivalent circulating density

- Pump barite or heavy weighting agents to plug the wellbore in order to stop flowing

- Pump cement to plug the wellbore

Shut-in Procedures

Using shut-in procedures is one of the oil-well-control measures to curtail kicks and prevent a blowout from occurring. Shut-in procedures are specific procedures for closing a well in case of a kick. When any positive indication of a kick is observed, such as a sudden increase in flow, or an increase in pit level, then the well should be shut-in immediately. If a well shut-in is not done promptly, a blowout is likely to happen.

Shut-in procedures are usually developed and practiced for every rig activity, such as drilling, tripping, logging, running tubular, performing a drill stem test, and so on. The primary purpose of a specific shut-in procedure is to minimize kick volume entering into a wellbore when a kick occurs, regardless of what phase of rig activity is occurring. However, a shut-in procedure is a company-specific procedure, and the policy of a company will dictate how a well should be shut-in.

They are generally two type of Shut-in procedures which are: soft shut-in, or hard shut-in.

Of these two methods, the hard shut-in is the fastest method to shut in the well; therefore, it will minimize the volume of kick allowed into the wellbore.

Well Kill Procedures

A well kill procedure is an oil well control method. Once the well has been shut-in on a kick, proper kill procedures must be done immediately. The general idea in well kill procedure is to circulate out any formation fluid already in the wellbore during kick, and then circulate a satisfactory weight of kill mud called Kill Weight Mud (KWM) into the well without allowing further fluid into the hole. If this can be done, then once the kill mud has been fully circulated around the well, it is possible to open up the well and restart normal operations. Generally, a kill mud (KWM) mix, which provides just hydrostatic balance for formation pressure, is circulated. This allows approximately constant bottom hole pressure, which is slightly greater than formation pressure to be maintained, as the kill circulation proceeds because of the additional small circulating friction pressure loss. After circulation, the well is opened up again.

The major well kill procedures used in oil well control are listed below:

- Wait and Weight
- Driller method
- Circulate and Weight
- Concurrent Method
- Reverse Circulation
- Dynamic Kill procedure
- Bullheading
- Volumetric Method
- Lubricate and Bleed

Oil well Control Incidents - Root Causes

There will always be potential oil well control problems, as long as there are drilling operations anywhere in the world. Most of these well control problems are as a result of some errors and can be eliminated, even though some are actually unavoidable. Since we know the consequences of failed well control are severe, efforts should be made to prevent some human errors which are the root causes of these incidents. These causes include:

- Lack of knowledge and skills of rig personnel

- Improper work practices

- Lack of understanding of oil well control training

- Lack of application of policies, procedures, and standards

- Inadequate risk management

A FOUNTAIN AT BIBI-EIBAT IN FLAMES, BAKU

1904 oil well fire at Bibi-Eibat (near Baku, Azerbaijan).

Organizations for Building Well-control Culture

Good oil-well-control culture requires personnel involved in oil well control to develop a core value for it by doing the proper thing at the proper time. A good well-control culture will definitely minimize well control incidents. Building well-control culture would involve developing competent personnel that are able to recognize well-control problems and know what to do to mitigate against them. This is usually done through quality-assurance programs and training. These programs are done by organizations such as the International Association of Drilling Contractors (IADC) or International Well Control Forum (IWCF).

IADC operates the Well-Control accreditation Program (WellCAP), which is a training program aimed at providing the necessary knowledge and practical skills critical to successful well control and to develop competent rig personnel. This training starts with floor-hand level and continues to the most-experienced drilling personnel.

IWCF is an NGO whose main aim is to develop and administer well-control certification programs for personnel employed in oil-well drilling, workover and well-intervention operations.

Oil Platform

Oil platform P-51 off the Brazilian coast is a semi-submersible platform.

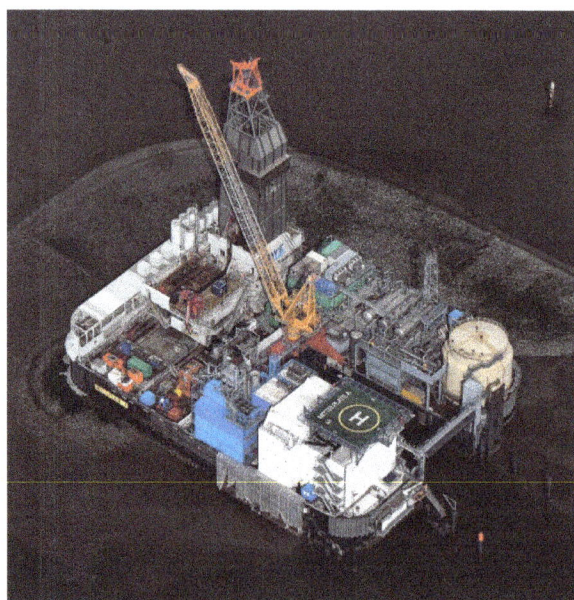

Oil platform Mittelplate in the North sea

An oil platform, offshore platform, or (colloquially and incorrectly) oil rig is a large structure with facilities to drill wells (optionally), to extract and process oil and natural gas, or to temporarily store product until it can be brought to shore for refining and marketing. In many cases, the platform contains facilities to house the workforce as well.

Depending on the circumstances, the platform may be fixed to the ocean floor, may consist of an artificial island, or may float. Remote subsea wells may also be connected to a platform by flow lines and by umbilical connections. These sub-sea solutions may consist of one or more subsea wells, or of one or more manifold centres for multiple wells.

History

Around 1891, the first submerged oil wells were drilled from platforms built on piles in the fresh waters of the Grand Lake St. Marys (a.k.a. Mercer County Reservoir) in Ohio. The wide but shallow reservoir was built from 1837 to 1845 to provide water to the Miami and Erie Canal.

Offshore platform, Gulf of Mexico

Around 1896, the first submerged oil wells in salt water were drilled in the portion of the Summerland field extending under the Santa Barbara Channel in California. The wells were drilled from piers extending from land out into the channel.

Other notable early submerged drilling activities occurred on the Canadian side of Lake Erie since 1913 and Caddo Lake in Louisiana in the 1910s. Shortly thereafter, wells were drilled in tidal zones along the Gulf Coast of Texas and Louisiana. The Goose Creek field near Baytown, Texas is one such example. In the 1920s, drilling was done from concrete platforms in Lake Maracaibo, Venezuela.

The oldest offshore well recorded in Infield's offshore database is the Bibi Eibat well which came on stream in 1923 in Azerbaijan. Landfill was used to raise shallow portions of the Caspian Sea.

In the early 1930s, the Texas Company developed the first mobile steel barges for drilling in the brackish coastal areas of the gulf.

In 1937, Pure Oil Company (now part of Chevron Corporation) and its partner Superior Oil Company (now part of ExxonMobil Corporation) used a fixed platform to develop a field in 14 feet (4.3 m) of water, one mile (1.6 km) offshore of Calcasieu Parish, Louisiana.

In 1946, Magnolia Petroleum Company (now part of ExxonMobil) erected a drilling platform in 18 ft (5.5 m) of water, 18 miles off the coast of St. Mary Parish, Louisiana.

In early 1947, Superior Oil erected a drilling/production platform in 20 ft (6.1 m) of water some 18 miles off Vermilion Parish, Louisiana. But it was Kerr-McGee Oil Industries (now Anadarko

Petroleum Corporation), as operator for partners Phillips Petroleum (ConocoPhillips) and Stanolind Oil & Gas (BP), that completed its historic Ship Shoal Block 32 well in October 1947, months before Superior actually drilled a discovery from their Vermilion platform farther offshore. In any case, that made Kerr-McGee's well the first oil discovery drilled out of sight of land.

The British Maunsell Forts constructed during World War II are considered the direct predecessors of modern offshore platforms. Having been pre-constructed in a very short time, they were then floated to their location and placed on the shallow bottom of the Thames and the Mersey estuary.

Types

Larger lake- and sea-based offshore platforms and drilling rig for oil.

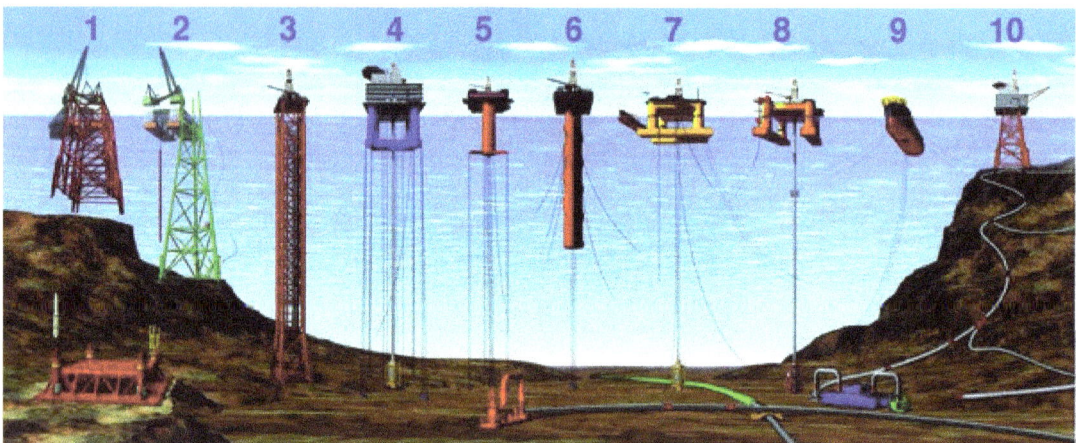

1, 2) conventional fixed platforms; 3) compliant tower; 4, 5) vertically moored tension leg and mini-tension leg platform; 6) spar; 7,8) semi-submersibles; 9) floating production, storage, and offloading facility; 10) sub-sea completion and tie-back to host facility.

Fixed Platforms

A fixed platform base under construction on a Louisiana river

These platforms are built on concrete or steel legs, or both, anchored directly onto the seabed, supporting a deck with space for drilling rigs, production facilities and crew quarters. Such platforms are, by virtue of their immobility, designed for very long term use (for instance the Hibernia platform). Various types of structure are used: steel jacket, concrete caisson, floating steel, and even floating concrete. Steel jackets are structural sections made of tubular steel members, and are usually piled into the seabed.

Concrete caisson structures, pioneered by the Condeep concept, often have in-built oil storage in tanks below the sea surface and these tanks were often used as a flotation capability, allowing them to be built close to shore (Norwegian fjords and Scottish firths are popular because they are sheltered and deep enough) and then floated to their final position where they are sunk to the seabed. Fixed platforms are economically feasible for installation in water depths up to about 520 m (1,710 ft).

Compliant Towers

These platforms consist of slender, flexible towers and a pile foundation supporting a conventional deck for drilling and production operations. Compliant towers are designed to sustain significant lateral deflections and forces, and are typically used in water depths ranging from 370 to 910 metres (1,210 to 2,990 ft).

Semi-submersible Platform

These platforms have hulls (columns and pontoons) of sufficient buoyancy to cause the structure to float, but of weight sufficient to keep the structure upright. Semi-submersible platforms can be moved from place to place and can be ballasted up or down by altering the amount of flooding in buoyancy tanks. They are generally anchored by combinations of chain, wire rope or polyester rope, or both, during drilling and/or production operations, though they can also be kept in place by the use of dynamic positioning. Semi-submersibles can be used in water depths from 60 to 3,000 metres (200 to 10,000 ft).

Jack-up Drilling Rigs

400 feet (120 m) tall jackup rig being towed by tugboats, Kachemak Bay, Alaska

Jack-up Mobile Drilling Units (or jack-ups), as the name suggests, are rigs that can be jacked up above the sea using legs that can be lowered, much like jacks. These MODUs (Mobile Offshore Drilling Units) are typically used in water depths up to 120 metres (390 ft), although some designs can go to 170 m (560 ft) depth. They are designed to move from place to place, and then anchor themselves by deploying the legs to the ocean bottom using a rack and pinion gear system on each leg.

Drillships

A drillship is a maritime vessel that has been fitted with drilling apparatus. It is most often used for exploratory drilling of new oil or gas wells in deep water but can also be used for scientific drilling. Early versions were built on a modified tanker hull, but purpose-built designs are used today. Most drillships are outfitted with a dynamic positioning system to maintain position over the well. They can drill in water depths up to 3,700 m (12,100 ft).

Floating Production Systems

The main types of floating production systems are FPSO (floating production, storage, and offloading system). FPSOs consist of large monohull structures, generally (but not always) shipshaped, equipped with processing facilities. These platforms are moored to a location for extended periods, and do not actually drill for oil or gas. Some variants of these applications, called FSO (floating storage and offloading system) or FSU (floating storage unit), are used exclusively for storage purposes, and host very little process equipment. This is one of the best sources for having floating production.

The world's first floating liquefied natural gas (FLNG) facility is currently under development.

Tension-leg Platform

TLPs are floating platforms tethered to the seabed in a manner that eliminates most vertical movement of the structure. TLPs are used in water depths up to about 2,000 meters (6,600 feet). The "conventional" TLP is a 4-column design which looks similar to a semisubmersible. Proprietary versions include the Seastar and MOSES mini TLPs; they are relatively low cost, used in water depths between 180 and 1,300 metres (590 and 4,270 ft). Mini TLPs can also be used as utility, satellite or early production platforms for larger deepwater discoveries.

Gravity-based Structure

A GBS can either be steel or concrete and is usually anchored directly onto the seabed. Steel GBS are predominantly used when there is no or limited availability of crane barges to install a conventional fixed offshore platform, for example in the Caspian Sea. There are several steel GBS in the world today (e.g. offshore Turkmenistan Waters (Caspian Sea) and offshore New Zealand). Steel GBS do not usually provide hydrocarbon storage capability. It is mainly installed by pulling it off the yard, by either wet-tow or/and dry-tow, and self-installing by controlled ballasting of the compartments with sea water. To position the GBS during installation, the GBS may be connected to either a transportation barge or any other barge (provided it is large enough to support the GBS) using strand jacks. The jacks shall be released gradually whilst the GBS is ballasted to ensure that the GBS does not sway too much from target location.

Spar Platforms

Spars are moored to the seabed like TLPs, but whereas a TLP has vertical tension tethers, a spar has more conventional mooring lines. Spars have to-date been designed in three configurations:

the "conventional" one-piece cylindrical hull; the "truss spar", in which the midsection is composed of truss elements connecting the upper buoyant hull (called a hard tank) with the bottom soft tank containing permanent ballast; and the "cell spar", which is built from multiple vertical cylinders. The spar has more inherent stability than a TLP since it has a large counterweight at the bottom and does not depend on the mooring to hold it upright. It also has the ability, by adjusting the mooring line tensions (using chain-jacks attached to the mooring lines), to move horizontally and to position itself over wells at some distance from the main platform location. The first production spar was Kerr-McGee's Neptune, anchored in 590 m (1,940 ft) in the Gulf of Mexico; however, spars (such as Brent Spar) were previously used as FSOs.

Devil's Tower spar platform

Eni's Devil's Tower located in 1,710 m (5,610 ft) of water in the Gulf of Mexico, was the world's deepest spar until 2010. The world's deepest platform is currently the Perdido spar in the Gulf of Mexico, floating in 2,438 metres of water. It is operated by Royal Dutch Shell and was built at a cost of $3 billion.

The first truss spars were Kerr-McGee's Boomvang and Nansen. The first (and only) cell spar is Kerr-McGee's Red Hawk.

Normally Unmanned Installations (NUI)

These installations, sometimes called toadstools, are small platforms, consisting of little more than a well bay, helipad and emergency shelter. They are designed to be operated remotely under normal conditions, only to be visited occasionally for routine maintenance or well work.

Conductor Support Systems

These installations, also known as satellite platforms, are small unmanned platforms consisting of little more than a well bay and a small process plant. They are designed to operate in conjunction with a static production platform which is connected to the platform by flow lines or by umbilical cable, or both.

Particularly Large Examples

A 'Statfjord' gravity-based structure under construction in Norway. Almost all of the structure will end up submerged.

The Petronius Platform is a compliant tower in the Gulf of Mexico modeled after the Hess Baldpate platform, which stands 2,000 feet (610 m) above the ocean floor. It is one of the world's tallest structures.

The Hibernia platform in Canada is the world's largest (in terms of weight) offshore platform, located on the Jeanne D'Arc Basin, in the Atlantic Ocean off the coast of Newfoundland. This *gravity base structure* (GBS), which sits on the ocean floor, is 111 metres (364 ft) high and has storage capacity for 1.3 million barrels (210,000 m^3) of crude oil in its 85-metre (279 ft) high caisson. The platform acts as a small concrete island with serrated outer edges designed to withstand the impact of an iceberg. The GBS contains production storage tanks and the remainder of the void space is filled with ballast with the entire structure weighing in at 1.2 million tons.

Royal Dutch Shell is currently developing the first Floating Liquefied Natural Gas (FLNG) facility, which will be situated approximately 200 km off the coast of Western Australia and is due for completion around 2017. When finished, it will be the largest floating offshore facility. It is expected to be approximately 488m long and 74m wide with displacement of around 600,000t when fully ballasted.

Maintenance and Supply

A typical oil production platform is self-sufficient in energy and water needs, housing electrical generation, water desalinators and all of the equipment necessary to process oil and gas such that it can be either delivered directly onshore by pipeline or to a floating platform or tanker loading facility, or both. Elements in the oil/gas production process include wellhead, production manifold, production separator, glycol process to dry gas, gas compressors, water injection pumps, oil/gas export metering and main oil line pumps.

Larger platforms assisted by smaller ESVs (emergency support vessels) like the British Iolair that are summoned when something has gone wrong, *e.g.* when a search and rescue operation is required. During normal operations, PSVs (platform supply vessels) keep the platforms provisioned and supplied, and AHTS vessels can also supply them, as well as tow them to location and serve as standby rescue and firefighting vessels.

Crew

Essential Personnel

Not all of the following personnel are present on every platform. On smaller platforms, one worker can perform a number of different jobs. The following also are not names officially recognized in the industry:

- OIM (offshore installation manager) who is the ultimate authority during his/her shift and makes the essential decisions regarding the operation of the platform;
- operations team leader (OTL);
- offshore operations engineer (OOE) who is the senior technical authority on the platform;
- PSTL or operations coordinator for managing crew changes;
- dynamic positioning operator, navigation, ship or vessel maneuvering (MODU), station keeping, fire and gas systems operations in the event of incident;
- automation systems specialist, to configure, maintain and troubleshoot the process control systems (PCS), process safety systems, emergency support systems and vessel management systems;
- second mate to meet manning requirements of flag state, operates fast rescue craft, cargo operations, fire team leader;
- third mate to meet manning requirements of flag state, operate fast rescue craft, cargo operations, fire team leader;
- ballast control operator to operate fire and gas systems;
- crane operators to operate the cranes for lifting cargo around the platform and between boats;
- scaffolders to rig up scaffolding for when it is required for workers to work at height;
- coxswains to maintain the lifeboats and manning them if necessary;
- control room operators, especially FPSO or production platforms;

- catering crew, including people tasked with performing essential functions such as cooking, laundry and cleaning the accommodation;

- production techs to run the production plant;

- helicopter pilot(s) living on some platforms that have a helicopter based offshore and transporting workers to other platforms or to shore on crew changes;

- maintenance technicians (instrument, electrical or mechanical).

- Fully qualified medic.

- Radio operator to operate all radio communications.

- Store Keeper, keeping the inventory well supplied

Incidental Personnel

Drill crew will be on board if the installation is performing drilling operations. A drill crew will normally comprise:

- Toolpusher
- Driller
- Roughnecks
- Roustabouts
- Company man
- Mud engineer
- Motorman
- Derrickhand
- Geologist
- Welders and Welder Helpers

Well services crew will be on board for well work. The crew will normally comprise:

- Well services supervisor
- Wireline or coiled tubing operators
- Pump operator
- Pump hanger and ranger

Drawbacks

Risks

The nature of their operation — extraction of volatile substances sometimes under extreme pressure in a hostile environment — means risk; accidents and tragedies occur regularly. The U.S. Minerals Management Service reported 69 offshore deaths, 1,349 injuries, and 858 fires and explosions on offshore rigs in the Gulf of Mexico from 2001 to 2010. On July 6, 1988, 167 people died when Occidental Petroleum's Piper Alpha offshore production platform, on the Piper field in the UK sector of the North Sea, exploded after a gas leak. The resulting investigation conducted by Lord Cullen and publicized in the first Cullen Report was highly critical of a number of areas, including, but not limited to, management within the company, the design of the structure, and the Permit to Work System. The report was commissioned in 1988, and was delivered November 1990. The accident greatly accelerated the practice of providing living accommodations on separate platforms, away from those used for extraction.

The offshore can be in itself a hazardous environment. In March 1980, the 'flotel' (floating hotel) platform *Alexander L. Kielland* capsized in a storm in the North Sea with the loss of 123 lives.

In 2001, *Petrobras 36* in Brazil exploded and sank five days later, killing 11 people.

Given the number of grievances and conspiracy theories that involve the oil business, and the importance of gas/oil platforms to the economy, platforms in the United States are believed to be potential terrorist targets. Agencies and military units responsible for maritime counter-terrorism in the US (Coast Guard, Navy SEALs, Marine Recon) often train for platform raids.

On April 21, 2010, the *Deepwater Horizon* platform, 52 miles off-shore of Venice, Louisiana, (property of Transocean and leased to BP) exploded, killing 11 people, and sank two days later. The resulting undersea gusher, conservatively estimated to exceed 20 million US gallons (76,000 m³) as of early June, 2010, became the worst oil spill in US history, eclipsing the Exxon Valdez oil spill.

Ecological Effects

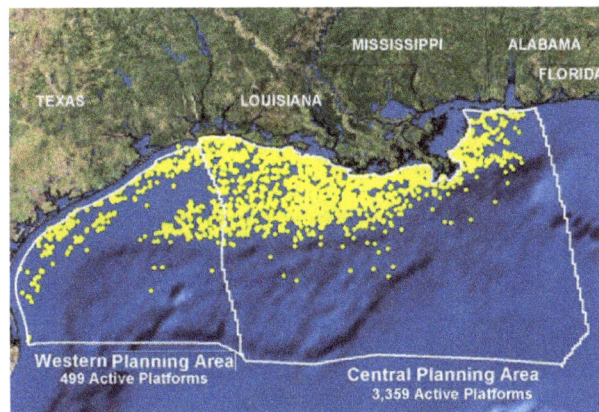

NOAA map of the 3,858 oil and gas platforms extant in the Gulf of Mexico in 2006

In British waters, the cost of removing all platform rig structures entirely was estimated in 2013 at £30 billion.

Aquatic organisms invariably attach themselves to the undersea portions of oil platforms, turning them into artificial reefs. In the Gulf of Mexico and offshore California, the waters around oil platforms are popular destinations for sports and commercial fishermen, because of the greater numbers of fish near the platforms. The United States and Brunei have active Rigs-to-Reefs programs, in which former oil platforms are left in the sea, either in place or towed to new locations, as permanent artificial reefs. In the US Gulf of Mexico, as of September 2012, 420 former oil platforms, about 10 percent of decommissioned platforms, have been converted to permanent reefs.

On the US Pacific coast, marine biologist Milton Love has proposed that oil platforms off California be retained as artificial reefs, instead of being dismantled (at great cost), because he has found them to be havens for many of the species of fish which are otherwise declining in the region, in the course of 11 years of research. Love is funded mainly by government agencies, but also in small part by the California Artificial Reef Enhancement Program. Divers have been used to assess the fish populations surrounding the platforms.

Deepest Oil Platforms

The world's deepest oil platform is the floating Perdido, which is a spar platform in the Gulf of Mexico in a water depth of 2,438 metres (7,999 ft).

Non-floating compliant towers and fixed platforms, by water depth:

- Petronius Platform, 535 m (1,755 ft)
- Baldpate Platform, 502 m (1,647 ft)
- Troll A Platform, 472 m (1,549 ft)
- Bullwinkle Platform, 413 m (1,355 ft)
- Pompano Platform, 393 m (1,289 ft)
- Benguela-Belize Lobito-Tomboco Platform, 390 m (1,280 ft)
- Gulfaks C Platform, 380 m (1,250 ft)
- Tombua Landana Platform, 366 m (1,201 ft)
- Harmony Platform, 366 m (1,201 ft)

Hydraulic Fracturing

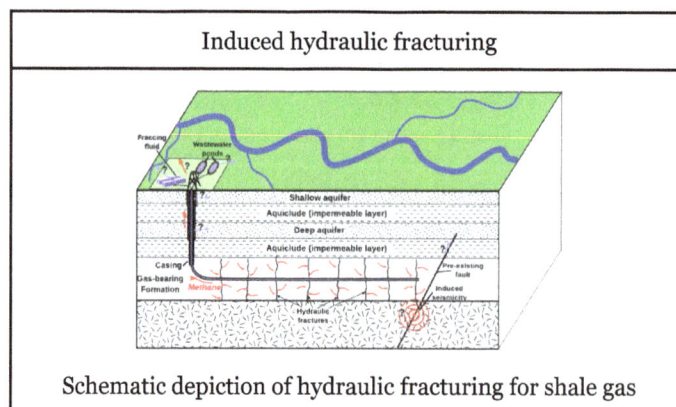

Schematic depiction of hydraulic fracturing for shale gas

Process type	Mechanical
Industrial sector(s)	Mining
Main technologies or sub-processes	Fluid pressure
Product(s)	Natural gas, petroleum
Inventor	Floyd Farris, Joseph B. Clark (Stanolind Oil and Gas Corporation)
Year of invention	1947

Hydraulic fracturing
Shale gas drilling rig near Alvarado, Texas
By country
• Canada • New Zealand • South Africa • United Kingdom • United States
Environmental impact
• Additives • United States

Regulation
• United States exemptions
Technology
• Proppants
• Uses of radioactivity
Politics
• 2012-14 Romanian protests
• Anti-fracking movement
• *FrackNation*
• Frack Off
• *Gasland*

Hydraulic fracturing (also fraccing, fracking, hydrofracturing or hydrofracking) is a well-stimulation technique in which rock is fractured by a pressurized liquid. The process involves the high-pressure injection of 'fracking fluid' (primarily water, containing sand or other proppants suspended with the aid of thickening agents) into a wellbore to create cracks in the deep-rock formations through which natural gas, petroleum, and brine will flow more freely. When the hydraulic pressure is removed from the well, small grains of hydraulic fracturing proppants (either sand or aluminium oxide) hold the fractures open.

Hydraulic fracturing began as an experiment in 1947, and the first commercially successful application followed in 1950. As of 2012, 2.5 million "frac jobs" had been performed worldwide on oil and gas wells; over one million of those within the U.S. Such treatment is generally necessary to achieve adequate flow rates in shale gas, tight gas, tight oil, and coal seam gas wells. Some hydraulic fractures can form naturally in certain veins or dikes.

Hydraulic fracturing is highly controversial in many countries. Its proponents advocate the economic benefits of more extensively accessible hydrocarbons. However, opponents argue that these are outweighed by the potential environmental impacts, which include risks of ground and surface water contamination, air and noise pollution, and the triggering of earthquakes, along with the consequential hazards to public health and the environment.

Increases in seismic activity following hydraulic fracturing along dormant or previously unknown faults are sometimes caused by the deep-injection disposal of hydraulic fracturing flowback (a byproduct of hydraulically fractured wells), and produced formation brine (a byproduct of both fractured and nonfractured oil and gas wells). For these reasons, hydraulic fracturing is under international scrutiny, restricted in some countries, and banned altogether in others. Some countries have banned the practice or put moratoria in place, while others have adopted an approach involving tight regulation. The European Union is drafting regulations that would permit the controlled application of hydraulic fracturing.

Geology

Halliburton fracturing operation in the Bakken Formation, North Dakota, United States

A fracturing operation in progress

Mechanics

Fracturing rocks at great depth frequently becomes suppressed by pressure due to the weight of the overlying rock strata and the cementation of the formation. This suppression process is particularly significant in "tensile" (Mode 1) fractures which require the walls of the fracture to move against this pressure. Fracturing occurs when effective stress is overcome by the pressure of fluids within the rock. The minimum principal stress becomes tensile and exceeds the tensile strength of the material. Fractures formed in this way are generally oriented in a plane perpendicular to the minimum principal stress, and for this reason, hydraulic fractures in well bores can be used to determine the orientation of stresses. In natural examples, such as dikes or vein-filled fractures, the orientations can be used to infer past states of stress.

Veins

Most mineral vein systems are a result of repeated natural fracturing during periods of relatively high pore fluid pressure. The impact of high pore fluid pressure on the formation process of mineral vein systems is particularly evident in "crack-seal" veins, where the vein material is part of a series of discrete fracturing events, and extra vein material is deposited on each occasion. One example of long-term repeated natural fracturing is in the effects of seismic activity. Stress levels rise and fall episodically, and earthquakes can cause large volumes of connate water to be expelled from fluid-filled fractures. This process is referred to as "seismic pumping".

Dikes

Minor intrusions in the upper part of the crust, such as dikes, propagate in the form of fluid-filled cracks. In such cases, the fluid is magma. In sedimentary rocks with a significant water content, fluid at fracture tip will be steam.

History

Precursors

Fracturing as a method to stimulate shallow, hard rock oil wells dates back to the 1860s. Dynamite or nitroglycerin detonations were used to increase oil and natural gas production from petroleum bearing formations. On 25 April 1865, Civil War veteran Col. Edward A. L. Roberts received a patent for an "exploding torpedo". It was employed in Pennsylvania, New York, Kentucky, and West Virginia using liquid and also, later, solidified nitroglycerin. Later still the same method was applied to water and gas wells. Stimulation of wells with acid, instead of explosive fluids, was introduced in the 1930s. Due to acid etching, fractures would not close completely resulting in further productivity increase.

Oil and Gas Wells

The relationship between well performance and treatment pressures was studied by Floyd Farris of Stanolind Oil and Gas Corporation. This study was the basis of the first hydraulic fracturing experiment, conducted in 1947 at the Hugoton gas field in Grant County of southwestern Kansas by Stanolind. For the well treatment, 1,000 US gallons (3,800 l; 830 imp gal) of gelled gasoline (essentially napalm) and sand from the Arkansas River was injected into the gas-producing limestone formation at 2,400 feet (730 m). The experiment was not very successful as deliverability of the well did not change appreciably. The process was further described by J.B. Clark of Stanolind in his paper published in 1948. A patent on this process was issued in 1949 and exclusive license was granted to the Halliburton Oil Well Cementing Company. On 17 March 1949, Halliburton performed the first two commercial hydraulic fracturing treatments in Stephens County, Oklahoma, and Archer County, Texas. Since then, hydraulic fracturing has been used to stimulate approximately one million oil and gas wells in various geologic regimes with good success.

In contrast with large-scale hydraulic fracturing used in low-permeability formations, small hydraulic fracturing treatments are commonly used in high-permeability formations to remedy "skin damage", a low-permeability zone that sometimes forms at the rock-borehole interface. In such cases the fracturing may extend only a few feet from the borehole.

In the Soviet Union, the first hydraulic proppant fracturing was carried out in 1952. Other countries in Europe and Northern Africa subsequently employed hydraulic fracturing techniques including Norway, Poland, Czechoslovakia, Yugoslavia, Hungary, Austria, France, Italy, Bulgaria, Romania, Turkey, Tunisia, and Algeria.

Massive Fracturing

Massive hydraulic fracturing (also known as high-volume hydraulic fracturing) is a technique first applied by Pan American Petroleum in Stephens County, Oklahoma, USA in 1968. The definition of massive hydraulic fracturing varies, but generally refers to treatments injecting over 150 short tons, or approximately 300,000 pounds (136 metric tonnes), of proppant.

Well head where fluids are injected into the ground

Well head after all the hydraulic fracturing equipment has been taken off location

American geologists gradually became aware that there were huge volumes of gas-saturated sandstones with permeability too low (generally less than 0.1 millidarcy) to recover the gas economically. Starting in 1973, massive hydraulic fracturing was used in thousands of gas wells in the San Juan Basin, Denver Basin, the Piceance Basin, and the Green River Basin, and in other hard rock formations of the western US. Other tight sandstone wells in the US made economically viable by massive hydraulic fracturing were in the Clinton-Medina Sandstone (Ohio, Pennsylvania, and New York), and Cotton Valley Sandstone (Texas and Louisiana).

Massive hydraulic fracturing quickly spread in the late 1970s to western Canada, Rotliegend and Carboniferous gas-bearing sandstones in Germany, Netherlands (onshore and offshore gas fields), and the United Kingdom in the North Sea.

Horizontal oil or gas wells were unusual until the late 1980s. Then, operators in Texas began completing thousands of oil wells by drilling horizontally in the Austin Chalk, and giving massive *slickwater* hydraulic fracturing treatments to the wellbores. Horizontal wells proved much more effective than vertical wells in producing oil from tight chalk; sedimentary beds are usually nearly horizontal, so horizontal wells have much larger contact areas with the target formation.

Shales

Hydraulic fracturing of shales goes back at least to 1965, when some operators in the Big Sandy gas field of eastern Kentucky and southern West Virginia started hydraulically fracturing the Ohio Shale and Cleveland Shale, using relatively small fracs. The frac jobs generally increased production, especially from lower-yielding wells.

In 1976, the United States government started the Eastern Gas Shales Project, which included numerous public-private hydraulic fracturing demonstration projects. During the same period, the Gas Research Institute, a gas industry research consortium, received approval for research and funding from the Federal Energy Regulatory Commission.

In 1997, Nick Steinsberger, an engineer of Mitchell Energy (now part of Devon Energy), applied the slickwater fracturing technique, using more water and higher pump pressure than previous fracturing techniques, which was used in East Texas by Union Pacific Resources (now part of Anadarko Petroleum Corporation), in the Barnett Shale of north Texas. In 1998, the new technique approved to be successful when the first 90 days gas production from the well called S.H. Griffin No. 3 exceeded production of any of the company's previous wells. This new completion technique made gas extraction widely economical in the Barnett Shale, and was later applied to other shales. George P. Mitchell has been called the "father of fracking" because of his role in applying it in shales. The first horizontal well in the Barnett Shale was drilled in 1991, but was not widely done in the Barnett until it was demonstrated that gas could be economically extracted from vertical wells in the Barnett.

As of 2013, massive hydraulic fracturing is being applied on a commercial scale to shales in the United States, Canada, and China. Several additional countries are planning to use hydraulic fracturing.

Process

According to the United States Environmental Protection Agency (EPA), hydraulic fracturing is a process to stimulate a natural gas, oil, or geothermal well to maximize extraction. The EPA defines the broader process to include acquisition of source water, well construction, well stimulation, and waste disposal.

Method

A hydraulic fracture is formed by pumping fracturing fluid into a wellbore at a rate sufficient to increase pressure at the target depth (determined by the location of the well casing perforations), to exceed that of the fracture *gradient* (pressure gradient) of the rock. The fracture gradient is defined as pressure increase per unit of depth relative to density, and is usually measured in pounds per square inch, per square foot, or bars. The rock cracks, and the fracture fluid permeates the rock extending the crack further, and further, and so on. Fractures are localized as pressure drops off with the rate of frictional loss, which is relevant to the distance from the well. Operators typically try to maintain "fracture width", or slow its decline following treatment, by introducing a proppant into the injected fluid – a material such as grains of sand, ceramic, or other particulate, thus preventing the fractures from closing when injection is stopped and pressure removed. Consideration of proppant strength and prevention of proppant failure becomes more important at greater depths where pressure and stresses on fractures are higher. The propped fracture is permeable enough to allow the flow of gas, oil, salt water and hydraulic fracturing fluids to the well.

During the process, fracturing fluid leakoff (loss of fracturing fluid from the fracture channel into the surrounding permeable rock) occurs. If not controlled, it can exceed 70% of the injected volume. This may result in formation matrix damage, adverse formation fluid interaction, and altered fracture geometry, thereby decreasing efficiency.

The location of one or more fractures along the length of the borehole is strictly controlled by various methods that create or seal holes in the side of the wellbore. Hydraulic fracturing is performed in cased wellbores, and the zones to be fractured are accessed by perforating the casing at those locations.

Hydraulic-fracturing equipment used in oil and natural gas fields usually consists of a slurry blender, one or more high-pressure, high-volume fracturing pumps (typically powerful triplex or quintuplex pumps) and a monitoring unit. Associated equipment includes fracturing tanks, one or more units for storage and handling of proppant, high-pressure treating iron, a chemical additive unit (used to accurately monitor chemical addition), low-pressure flexible hoses, and many gauges and meters for flow rate, fluid density, and treating pressure. Chemical additives are typically 0.5% percent of the total fluid volume. Fracturing equipment operates over a range of pressures and injection rates, and can reach up to 100 megapascals (15,000 psi) and 265 litres per second (9.4 cu ft/s) (100 barrels per minute).

Well Types

A distinction can be made between conventional, low-volume hydraulic fracturing, used to stimulate high-permeability reservoirs for a single well, and unconventional, high-volume hydraulic fracturing, used in the completion of tight gas and shale gas wells. High-volume hydraulic fracturing usually requires higher pressures than low-volume fracturing; the higher pressures are needed to push out larger volumes of fluid and proppant that extend farther from the borehole.

Horizontal drilling involves wellbores with a terminal drillhole completed as a "lateral" that extends parallel with the rock layer containing the substance to be extracted. For example, laterals extend 1,500 to 5,000 feet (460 to 1,520 m) in the Barnett Shale basin in Texas, and up to 10,000 feet (3,000 m) in the Bakken formation in North Dakota. In contrast, a vertical well only accesses the thickness of the rock layer, typically 50–300 feet (15–91 m). Horizontal drilling reduces surface disruptions as fewer wells are required to access the same volume of rock.

Drilling often plugs up the pore spaces at the wellbore wall, reducing permeability at and near the wellbore. This reduces flow into the borehole from the surrounding rock formation, and partially seals off the borehole from the surrounding rock. Low-volume hydraulic fracturing can be used to restore permeability.

Fracturing Fluids

The main purposes of fracturing fluid are to extend fractures, add lubrication, change gel strength, and to carry proppant into the formation. There are two methods of transporting proppant in the fluid – high-rate and high-viscosity. High-viscosity fracturing tends to cause large dominant fractures, while high-rate (slickwater) fracturing causes small spread-out micro-fractures.

Water tanks preparing for hydraulic fracturing

Water-soluble gelling agents (such as guar gum) increase viscosity and efficiently deliver proppant into the formation.

Process of mixing water with hydraulic fracturing fluids to be injected into the ground

Fluid is typically a slurry of water, proppant, and chemical additives. Additionally, gels, foams, and compressed gases, including nitrogen, carbon dioxide and air can be injected. Typically, 90% of the fluid is water and 9.5% is sand with chemical additives accounting to about 0.5%. However, fracturing fluids have been developed using liquefied petroleum gas (LPG) and propane in which water is unnecessary.

The proppant is a granular material that prevents the created fractures from closing after the fracturing treatment. Types of proppant include silica sand, resin-coated sand, bauxite, and man-made ceramics. The choice of proppant depends on the type of permeability or grain strength needed. In some formations, where the pressure is great enough to crush grains of natural silica sand, higher-strength proppants such as bauxite or ceramics may be used. The most commonly used proppant is silica sand, though proppants of uniform size and shape, such as a ceramic proppant, are believed to be more effective.

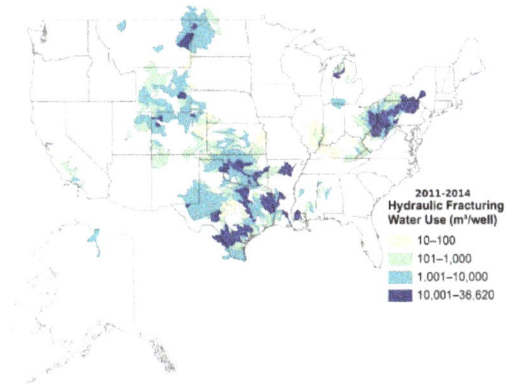

USGS map of water use from hydraulic fracturing between 2011 and 2014.
One cubic meter of water is 264.172 gallons.

The fracturing fluid varies depending on fracturing type desired, and the conditions of specific wells being fractured, and water characteristics. The fluid can be gel, foam, or slickwater-based. Fluid choices are tradeoffs: more viscous fluids, such as gels, are better at keeping proppant in suspension; while less-viscous and lower-friction fluids, such as slickwater, allow fluid to be pumped at higher rates, to create fractures farther out from the wellbore. Important material properties of the fluid include viscosity, pH, various rheological factors, and others.

Water is mixed with sand and chemicals to create fracking fluid. Approximately 40,000 gallons of chemicals are used per fracturing. A typical fracture treatment uses between 3 and 12 additive chemicals. Although there may be unconventional fracturing fluids, typical chemical additives can include one or more of the following:

- Acids—hydrochloric acid or acetic acid is used in the pre-fracturing stage for cleaning the perforations and initiating fissure in the near-wellbore rock.

- Sodium chloride (salt)—delays breakdown of gel polymer chains.

- Polyacrylamide and other friction reducers decrease turbulence in fluid flow and pipe friction, thus allowing the pumps to pump at a higher rate without having greater pressure on the surface.

- Ethylene glycol—prevents formation of scale deposits in the pipe.

- Borate salts—used for maintaining fluid viscosity during the temperature increase.

- Sodium and potassium carbonates—used for maintaining effectiveness of crosslinkers.

- Anaerobic, Biocide, BIO—Glutaraldehyde used as disinfectant of the water (bacteria elimination).

- Guar gum and other water-soluble gelling agents—increases viscosity of the fracturing fluid to deliver proppant into the formation more efficiently.

- Citric acid—used for corrosion prevention.

- Isopropanol—used to winterize the chemicals to ensure it doesn't freeze.

The most common chemical used for hydraulic fracturing in the United States in 2005–2009 was methanol, while some other most widely used chemicals were isopropyl alcohol, 2-butoxyethanol, and ethylene glycol.

Typical fluid types are:

- Conventional linear gels. These gels are cellulose derivative (carboxymethyl cellulose, hydroxyethyl cellulose, carboxymethyl hydroxyethyl cellulose, hydroxypropyl cellulose, hydroxyethyl methyl cellulose), guar or its derivatives (hydroxypropyl guar, carboxymethyl hydroxypropyl guar), mixed with other chemicals.

- Borate-crosslinked fluids. These are guar-based fluids cross-linked with boron ions (from aqueous borax/boric acid solution). These gels have higher viscosity at pH 9 onwards and are used to carry proppant. After the fracturing job, the pH is reduced to 3–4 so that the cross-links are broken, and the gel is less viscous and can be pumped out.

- Organometallic-crosslinked fluids - zirconium, chromium, antimony, titanium salts - are known to crosslink guar-based gels. The crosslinking mechanism is not reversible, so once the proppant is pumped down along with cross-linked gel, the fracturing part is done. The gels are broken down with appropriate breakers.

- Aluminium phosphate-ester oil gels. Aluminium phosphate and ester oils are slurried to form cross-linked gel. These are one of the first known gelling systems.

For slickwater fluids the use of sweeps is common. Sweeps are temporary reductions in the proppant concentration, which help ensure that the well is not overwhelmed with proppant. As the fracturing process proceeds, viscosity-reducing agents such as oxidizers and enzyme breakers are sometimes added to the fracturing fluid to deactivate the gelling agents and encourage flowback. Such oxidizer react with and break down the gel, reducing the fluid's viscosity and ensuring that no proppant is pulled from the formation. An enzyme acts as a catalyst for breaking down the gel. Sometimes pH modifiers are used to break down the crosslink at the end of a hydraulic fracturing job, since many require a pH buffer system to stay viscous. At the end of the job, the well is commonly flushed with water under pressure (sometimes blended with a friction reducing chemical.) Some (but not all) injected fluid is recovered. This fluid is managed by several methods, including underground injection control, treatment, discharge, recycling, and temporary storage in pits or containers. New technology is continually developing to better handle waste water and improve re-usability.

Fracture Monitoring

Measurements of the pressure and rate during the growth of a hydraulic fracture, with knowledge of fluid properties and proppant being injected into the well, provides the most common and simplest method of monitoring a hydraulic fracture treatment. This data along with knowledge of the underground geology can be used to model information such as length, width and conductivity of a propped fracture.

Injection of radioactive tracers along with the fracturing fluid is sometimes used to determine the injection profile and location of created fractures. Radiotracers are selected to have the readily detectable radiation, appropriate chemical properties, and a half life and toxicity level that will minimize initial and residual contamination. Radioactive isotopes chemically bonded to glass (sand) and/or resin beads may also be injected to track fractures. For example, plastic pellets coated with 10 GBq of Ag-110mm may be added to the proppant, or sand may be labelled with Ir-192, so that the proppant's progress can be monitored. Radiotracers such as Tc-99m and I-131 are also used to measure flow rates. The Nuclear Regulatory Commission publishes guidelines which list a wide range of radioactive materials in solid, liquid and gaseous forms that may be used as tracers and limit the amount that may be used per injection and per well of each radionuclide.

A new technique in well-monitoring involves fiber-optic cables outside the casing. Using the fiber optics, temperatures can be measured every foot along the well - even while the wells are being fracked and pumped. By monitoring the temperature of the well, engineers can determine how much fracking fluid different parts of the well use as well as how much natural gas they collect.

Microseismic Monitoring

For more advanced applications, microseismic monitoring is sometimes used to estimate the size and orientation of induced fractures. Microseismic activity is measured by placing an array of geophones in a nearby wellbore. By mapping the location of any small seismic events associated with the growing fracture, the approximate geometry of the fracture is inferred. Tiltmeter arrays deployed on the surface or down a well provide another technology for monitoring strain

Microseismic mapping is very similar geophysically to seismology. In earthquake seismology, seismometers scattered on or near the surface of the earth record S-waves and P-waves that are released during an earthquake event. This allows for motion along the fault plane to be estimated and its location in the earth's subsurface mapped. Hydraulic fracturing, an increase in formation stress proportional to the net fracturing pressure, as well as an increase in pore pressure due to leakoff. Tensile stresses are generated ahead of the fracture's tip, generating large amounts of shear stress. The increases in pore water pressure and in formation stress combine and affect weaknesses near the hydraulic fracture, like natural fractures, joints, and bedding planes.

Different methods have different location errors and advantages. Accuracy of microseismic event mapping is dependent on the signal-to-noise ratio and the distribution of sensors. Accuracy of events located by seismic inversion is improved by sensors placed in multiple azimuths from the monitored borehole. In a downhole array location, accuracy of events is improved by being close to the monitored borehole (high signal-to-noise ratio).

Monitoring of microseismic events induced by reservoir stimulation has become a key aspect in evaluation of hydraulic fractures, and their optimization. The main goal of hydraulic fracture monitoring is to completely characterize the induced fracture structure, and distribution of conductivity within a formation. Geomechanical analysis, such as understanding a formations material properties, in-situ conditions, and geometries, helps monitoring by providing a better definition of the environment in which the fracture network propagates. The next task is to know the location of proppant within the fracture and the distribution of fracture conductivity. This can be monitored using multiple types of techniques to finally develop a reservoir model than accurately predicts well performance.

Horizontal Completions

Since the early 2000s, advances in drilling and completion technology have made horizontal well-bores much more economical. Horizontal wellbores allow far greater exposure to a formation than conventional vertical wellbores. This is particularly useful in shale formations which do not have sufficient permeability to produce economically with a vertical well. Such wells, when drilled on-shore, are now usually hydraulically fractured in a number of stages, especially in North America. The type of wellbore completion is used to determine how many times a formation is fractured, and at what locations along the horizontal section.

In North America, shale reservoirs such as the Bakken, Barnett, Montney, Haynesville, Marcellus, and most recently the Eagle Ford, Niobrara and Utica shales are drilled horizontally through the producing interval(s), completed and fractured. The method by which the fractures are placed along the wellbore is most commonly achieved by one of two methods, known as "plug and perf" and "sliding sleeve".

The wellbore for a plug-and-perf job is generally composed of standard steel casing, cemented or uncemented, set in the drilled hole. Once the drilling rig has been removed, a wireline truck is used to perforate near the bottom of the well, and then fracturing fluid is pumped. Then the wireline truck sets a plug in the well to temporarily seal off that section so the next section of the wellbore can be treated. Another stage is pumped, and the process is repeated along the horizontal length of the wellbore.

The wellbore for the sliding sleeve technique is different in that the sliding sleeves are included at set spacings in the steel casing at the time it is set in place. The sliding sleeves are usually all closed at this time. When the well is due to be fractured, the bottom sliding sleeve is opened using one of several activation techniques and the first stage gets pumped. Once finished, the next sleeve is opened, concurrently isolating the previous stage, and the process repeats. For the sliding sleeve method, wireline is usually not required.

Sleeves

These completion techniques may allow for more than 30 stages to be pumped into the horizontal section of a single well if required, which is far more than would typically be pumped into a vertical well that had far fewer feet of producing zone exposed.

Uses

Hydraulic fracturing is used to increase the rate at which fluids, such as petroleum, water, or natural gas can be recovered from subterranean natural reservoirs. Reservoirs are typically porous sandstones, limestones or dolomite rocks, but also include "unconventional reservoirs" such as shale rock or coal beds. Hydraulic fracturing enables the extraction of natural gas and oil from rock formations deep below the earth's surface (generally 2,000–6,000 m (5,000–20,000 ft)), which is greatly below typical groundwater reservoir levels. At such depth, there may be insufficient permeability or reservoir pressure to allow natural gas and oil to flow from the rock into the wellbore at high economic return. Thus, creating conductive fractures in the rock is instrumental in extraction from naturally impermeable shale reservoirs. Permeability is measured in the microdarcy to nanodarcy range. Fractures are a conductive path connecting a larger volume of reservoir to the well. So-called "super fracking," creates cracks deeper in the rock formation to release more oil and gas, and increases efficiency. The yield for typical shale bores generally falls off after the first year or two, but the peak producing life of a well can be extended to several decades.

While the main industrial use of hydraulic fracturing is in stimulating production from oil and gas wells, hydraulic fracturing is also applied:

- To stimulate groundwater wells
- To precondition or induce rock cave-ins mining
- As a means of enhancing waste remediation, usually hydrocarbon waste or spills
- To dispose waste by injection deep into rock
- To measure stress in the Earth
- For electricity generation in enhanced geothermal systems
- To increase injection rates for geologic sequestration of CO_2

Since the late 1970s, hydraulic fracturing has been used, in some cases, to increase the yield of drinking water from wells in a number of countries, including the US, Australia, and South Africa.

Economic Effects

Hydraulic fracturing has been seen as one of the key methods of extracting unconventional oil and unconventional gas resources. According to the International Energy Agency, the remaining technically recoverable resources of shale gas are estimated to amount to 208 trillion cubic metres (7,300 trillion cubic feet), tight gas to 76 trillion cubic metres (2,700 trillion cubic feet), and coalbed methane to 47 trillion cubic metres (1,700 trillion cubic feet). As a rule, formations of these resources have lower permeability than conventional gas formations. Therefore, depending on the geological characteristics of the formation, specific technologies such as hydraulic fracturing are required. Although there are also other methods to extract these resources, such as conventional drilling or horizontal drilling, hydraulic fracturing is one of the key methods making their extraction economically viable. The multi-stage fracturing technique has facilitated the development of shale gas and light tight oil production in the United States and is believed to do so in the other countries with unconventional hydrocarbon resources.

Some studies call into question the claim that hydraulic fracturing of shale gas wells has a significant macro-economic impact. A study released in the beginning of 2014 by the Institute for Sustainable Development and International Relations states that, on the long-term as well as on the short-run, the "shale gas revolution" due to hydraulic fracturing in the United States has had very little impact on economic growth and competitiveness. The same report concludes that in Europe, using hydraulic fracturing would have very little advantage in terms of competitiveness and energy security. Indeed, for the period 2030-2035, shale gas is estimated to cover 3 to 10% of EU projected energy demand, which is not enough to have a significant impact on energetic independence and competitiveness.

Public Debate

Poster against hydraulic fracturing in Vitoria-Gasteiz, Spain, October 2012

Politics and Public Policy

An anti-fracking movement has emerged both internationally with involvement of international environmental organizations and nations such as France and locally in affected areas such as Balcombe in Sussex where the Balcombe drilling protest was in progress during mid-2013. The considerable opposition against hydraulic fracturing activities in local townships in the United States has led companies to adopt a variety of public relations measures to reassure the public, including the employment of former military personnel with training in psychological warfare operations. According to Matt Pitzarella, the communications director at Range Resources, employees trained in the Middle East have been valuable to Range Resources in Pennsylvania, when dealing with emotionally charged township meetings and advising townships on zoning and local ordinances dealing with hydraulic fracturing.

There have been many protests directed at hydraulic fracturing. For example, ten people were arrested in 2013 during an anti-fracking protest near New Matamoras, Ohio, after they illegally entered a development zone and latched themselves to drilling equipment. Though usually non-violent, some protestors use violence and intimidation. In northwest Pennsylvania, there was a drive-by shooting at a well site, in which someone shot two rounds of a small-caliber rifle in the direction of a drilling rig, before shouting profanities at the site and fleeing the scene. In Washington County, Pennsylvania, a contractor working on a gas pipeline found a pipe bomb that had been placed where a pipeline was to be constructed, which local authorities said would have caused a "catastrophe" had they not discovered and detonated it.

In 2014 a number of European officials suggested that several major European protests against fracking (with mixed success in Lithuania and Ukraine) may be partially sponsored by Gazprom, Russia's state-controlled gas company. The New York Times suggested that Russia saw its natural gas exports to Europe as a key element of its geopolitical influence, and that this market would diminish if fracking is adopted in Eastern Europe, as it opens up significant shale gas reserves in the region. Russian officials have on numerous occasions made public statements to the effect that fracking "poses a huge environmental problem".

Documentary Films

Josh Fox's 2010 Academy Award nominated film *Gasland* became a center of opposition to hydraulic fracturing of shale. The movie presented problems with groundwater contamination near well sites in Pennsylvania, Wyoming, and Colorado. *Energy in Depth*, an oil and gas industry lobbying group, called the film's facts into question. In response, a rebuttal of *Energy in Depth's* claims of inaccuracy was posted on *Gasland's* website.

The Director of the Colorado Oil and Gas Conservation Commission (COGCC) offered to be interviewed as part of the film if he could review what was included from the interview in the final film but Fox declined the offer. Exxon Mobil, Chevron Corporation and ConocoPhillips aired advertisements during 2011 and 2012 that claimed to describe the economic and environmental benefits of natural gas and argue that hydraulic fracturing was safe.

The film *Promised Land*, starring Matt Damon, takes on hydraulic fracturing. The gas industry is making plans to try to counter the film's criticisms of hydraulic fracturing with informational flyers, and Twitter and Facebook posts.

In January 2013 Northern Irish journalist and filmmaker Phelim McAleer released a crowdfunded documentary called *FrackNation* as a response to the statements made by Fox in *Gasland*. *FrackNation* premiered on Mark Cuban's AXS TV. The premiere corresponded with the release of *Promised Land*.

In April 2013, Josh Fox released *Gasland 2*, a documentary that states that the gas industry's portrayal of natural gas as a clean and safe alternative to oil is a myth, and that hydraulically fractured wells inevitably leak over time, contaminating water and air, hurting families, and endangering the earth's climate with the potent greenhouse gas methane.

In 2014, Vido Innovations released the documentary *The Ethics of Fracking*. The film covers the politics, spiritual, scientific, medical and professional points of view on hydraulic fracturing. It also digs into the way the gas industry portrays fracking in their advertising.

In 2015, the Canadian documentary film *Fractured Land* had its world premiere at the Hot Docs Canadian International Documentary Festival.

Research Issues

Typically the funding source of the research studies is a focal point of controversy. Concerns have been raised about research funded by foundations and corporations, or by environmental groups, which can at times lead to at least the appearance of unreliable studies. Several organizations, researchers, and media outlets have reported difficulty in conducting and reporting the results of studies on hydraulic fracturing due to industry and governmental pressure, and expressed concern over possible censoring of environmental reports. Some have argued there is a need for more research into the environmental and health effects of the technique.

However, it is important to note that many of the most-cited studies over the last decade are either government estimates, environmentalist group reports, or peer-reviewed papers from academic scientists, including a 2013 EPA report that significantly lowered estimates of methane leakage compared to previous estimates, and a study commissioned by the environmentalist group Environmental Defense Fund and published in the Proceedings of the National Academy of Sciences, which similarly showed that the environmental effects of natural gas production were overestimated. Similarly, a study from the environmentalist Natural Resources Defense Council was cited to show that a previous highly cited study "significantly overestimate[s] the fugitive emissions associated with unconventional gas extraction."

Health Risks

There is concern over the possible adverse public health implications of hydraulic fracturing activity. A 2013 review on shale gas production in the United States stated, "with increasing numbers of drilling sites, more people are at risk from accidents and exposure to harmful substances used at fractured wells." A 2011 hazard assessment recommended full disclosure of chemicals used for hydraulic fracturing and drilling as many have immediate health effects, and many may have long-term health effects.

In June 2014 Public Health England published a review of the potential public health impacts of exposures to chemical and radioactive pollutants as a result of shale gas extraction in the UK, based on the examination of literature and data from countries where hydraulic fracturing already occurs. The executive summary of the report stated: "An assessment of the currently available evidence indicates that the potential risks to public health from exposure to the emissions associated with shale gas extraction will be low if the operations are properly run and regulated. Most evidence suggests that contamination of groundwater, if it occurs, is most likely to be caused by leakage through the vertical borehole. Contamination of groundwater from the underground hydraulic fracturing process itself (ie the fracturing of the shale) is unlikely. However, surface spills of hydraulic fracturing fluids or wastewater may affect groundwater, and emissions to air also have the potential to impact on health. Where potential risks have been identified in the literature, the reported problems are typically a result of operational failure and a poor regulatory environment."

A 2012 report prepared for the European Union Directorate-General for the Environment identified potential risks to humans from air pollution and ground water contamination posed by hydraulic fracturing. This led to a series of recommendations in 2014 to mitigate these concerns. A 2012 guidance for pediatric nurses in the US said that hydraulic fracturing had a potential negative impact on public health and that pediatric nurses should be prepared to gather information on such topics so as to advocate for improved community health.

Environmental Impacts

The potential environmental impacts of hydraulic fracturing include air emissions and climate change, high water consumption, water contamination, land use, risk of earthquakes, noise pollution, and health effects on humans. Air emissions are primarily methane that escapes from wells, along with industrial emissions from equipment used in the extraction process. Modern UK and EU regulation requires zero emissions of methane, a potent greenhouse gas. Escape of methane is a bigger problem in older wells than in ones built under more recent EU legislation.

Hydraulic fracturing uses between 1.2 and 3.5 million US gallons (4,500 and 13,200 m³) of water per well, with large projects using up to 5 million US gallons (19,000 m³). Additional water is used when wells are refractured. An average well requires 3 to 8 million US gallons (11,000 to 30,000 m³) of water over its lifetime. According to the Oxford Institute for Energy Studies, greater volumes of fracturing fluids are required in Europe, where the shale depths average 1.5 times greater than in the U.S. Surface water may be contaminated through spillage and improperly built and maintained waste pits, and ground water can be contaminated if the fluid is able to escape the formation being fractured (through, for example, abandoned wells) or by produced water (the returning fluids, which also contain dissolved constituents such as minerals and brine waters). Produced water is managed by underground injection, municipal and commercial wastewater treatment and discharge, self-contained systems at well sites or fields, and recycling to fracture future wells. Typically less than half of the produced water used to fracture the formation is recovered.

About 3.6 hectares (8.9 acres) of land is needed per each drill pad for surface installations. Well pad and supporting structure construction significantly fragments landscapes which likely has negative effects on wildlife. These sites need to be remediated after wells are exhausted. Each well pad (in average 10 wells per pad) needs during preparatory and hydraulic fracturing process about 800 to 2,500 days of noisy activity, which affect both residents and local wildlife. In addition, noise is created by continuous truck traffic (sand, etc.) needed in hydraulic fracturing. Research is underway to determine if human health has been affected by air and water pollution, and rigorous following of safety procedures and regulation is required to avoid harm and to manage the risk of accidents that could cause harm.

In July 2013, the US Federal Railroad Administration listed oil contamination by hydraulic fracturing chemicals as "a possible cause" of corrosion in oil tank cars.

Hydraulic fracturing sometimes causes induced seismicity or earthquakes. The magnitude of these events is usually too small to be detected at the surface, although tremors attributed to fluid injection into disposal wells have been large enough to have often been felt by people, and to have caused property damage and possibly injuries.

Microseismic events are often used to map the horizontal and vertical extent of the fracturing. A better understanding of the geology of the area being fracked and used for injection wells can be helpful in mitigating the potential for significant seismic events.

Regulations

Countries using or considering use of hydraulic fracturing have implemented different regulations, including developing federal and regional legislation, and local zoning limitations. In 2011, after public pressure France became the first nation to ban hydraulic fracturing, based on the precautionary principle as well as the principle of preventive and corrective action of environmental hazards. The ban was upheld by an October 2013 ruling of the Constitutional Council. Some other countries such as Scotland have placed a temporary moratorium on the practice due to public health concerns and strong public opposition. Countries like the United Kingdom and South Africa have lifted their bans, choosing to focus on regulation instead of outright prohibition. Germany has announced draft regulations that would allow using hydraulic fracturing for the exploitation of shale gas deposits with the exception of wetland areas. In China, regulation on shale gas still faces hurdles, as it has complex interrelations with other regulatory regimes, especially trade.

The European Union has adopted a recommendation for minimum principles for using high-volume hydraulic fracturing. Its regulatory regime requires full disclosure of all additives. In the United States, the Ground Water Protection Council launched FracFocus.org, an online voluntary disclosure database for hydraulic fracturing fluids funded by oil and gas trade groups and the U.S. Department of Energy. Hydraulic fracturing is excluded from the Safe Drinking Water Act's underground injection control's regulation, except when diesel fuel is used. The EPA assures surveillance of the issuance of drilling permits when diesel fuel is employed.

In 2012, Vermont became the first state in the United States to ban hydraulic fracturing. On 17 December 2014, New York became the second state to issue a complete ban on any hydraulic fracturing due to potential risks to human health and the environment.

Offshore Drilling

An oil drilling platform off the coast of Santa Barbara, CA - 6 December 2011

Offshore drilling is a mechanical process where a wellbore is drilled below the seabed. It is typically carried out in order to explore for and subsequently extract petroleum which lies in rock formations beneath the seabed. Most commonly, the term is used to describe drilling activities on the continental shelf, though the term can also be applied to drilling in lakes, inshore waters and inland seas.

Offshore drilling presents environmental challenges, both from the produced hydrocarbons and the materials used during the drilling operation. Controversies include the ongoing US offshore drilling debate.

There are many different types of facilities from which offshore drilling operations take place. These include bottom founded drilling rigs (jackup barges and swamp barges), combined drilling and production facilities either bottom founded or floating platforms, and deepwater mobile offshore drilling units (MODU) including semi-submersibles and drillships. These are capable of operating in water depths up to 3,000 metres (9,800 ft). In shallower waters the mobile units are anchored to the seabed, however in deeper water (more than 1,500 metres (4,900 ft) the semisubmersibles or drillships are maintained at the required drilling location using dynamic positioning.

History

Around 1891, the first submerged oil wells were drilled from platforms built on piles in the fresh waters of the Grand Lake St. Marys (a.k.a. Mercer County Reservoir) in Ohio. The wells were developed by small local companies such as Bryson, Riley Oil, German-American and Banker's Oil.

Around 1896, the first submerged oil wells in salt water were drilled in the portion of the Summerland field extending under the Santa Barbara Channel in California. The wells were drilled from piers extending from land out into the channel.

Other notable early submerged drilling activities occurred on the Canadian side of Lake Erie in the 1900s and Caddo Lake in Louisiana in the 1910s. Shortly thereafter wells were drilled in tidal zones along the Texas and Louisiana gulf coast. The Goose Creek Oil Field near Baytown, Texas is one such example. In the 1920s drilling activities occurred from concrete platforms in Venezuela's Lake Maracaibo.

One of the oldest subsea wells is the Bibi Eibat well, which came on stream in 1923 in Azerbaijan. The well was located on an artificial island in a shallow portion of the Caspian Sea. In the early 1930s, the Texas Co., later Texaco (now Chevron) developed the first mobile steel barges for drilling in the brackish coastal areas of the Gulf of Mexico.

In 1937, Pure Oil (now Chevron) and its partner Superior Oil (now ExxonMobil) used a fixed platform to develop a field 1 mile (1.6 km) offshore of Calcasieu Parish, Louisiana in 14 feet (4.3 m) of water.

In 1945, concern for American control of its offshore oil reserves caused President Harry Truman to issue an Executive Order unilaterally extending American territory to the edge of its continental shelf, an act that effectively ended the 3-mile limit "freedom of the seas" regime.

In 1946, Magnolia Petroleum (now ExxonMobil) drilled at a site 18 miles (29 km) off the coast, erecting a platform in 18 feet (5.5 m) of water off St. Mary Parish, Louisiana.

In early 1947, Superior Oil erected a drilling and production platform in 20 feet (6.1 m) of water some 18 miles (29 km) off Vermilion Parish, La. But it was Kerr-McGee Oil Industries (now Anadarko Petroleum), as operator for partners Phillips Petroleum (ConocoPhillips) and Stanolind Oil & Gas (BP) that completed its historic Ship Shoal Block 32 well in October 1947, months before

Superior actually drilled a discovery from their Vermilion platform farther offshore. In any case, that made Kerr-McGee's well the first oil discovery drilled out of sight of land.

When offshore drilling moved into deeper waters of up to 30 metres (98 ft), fixed platform rigs were built, until demands for drilling equipment was needed in the 100 feet (30 m) to 120 metres (390 ft) depth of the Gulf of Mexico, the first jack-up rigs began appearing from specialized off-shore drilling contractors such as forerunners of ENSCO International.

The first semi-submersible resulted from an unexpected observation in 1961. Blue Water Drilling Company owned and operated the four-column submersible Blue Water Rig No.1 in the Gulf of Mexico for Shell Oil Company. As the pontoons were not sufficiently buoyant to support the weight of the rig and its consumables, it was towed between locations at a draught midway between the top of the pontoons and the underside of the deck. It was noticed that the motions at this draught were very small, and Blue Water Drilling and Shell jointly decided to try operating the rig in the floating mode. The concept of an anchored, stable floating deep-sea platform had been designed and tested back in the 1920s by Edward Robert Armstrong for the purpose of operating aircraft with an invention known as the 'seadrome'. The first purpose-built drilling semi-submersible *Ocean Driller* was launched in 1963. Since then, many semi-submersibles have been purpose-designed for the drilling industry mobile offshore fleet.

The first offshore drillship was the *CUSS 1* developed for the Mohole project to drill into the Earth's crust.

As of June, 2010, there were over 620 mobile offshore drilling rigs (Jackups, semisubs, drillships, barges) available for service in the competitive rig fleet.

One of the world's deepest hubs is currently the Perdido in the Gulf of Mexico, floating in 2,438 meters of water. It is operated by Royal Dutch Shell and was built at a cost of $3 billion. The deepest operational platform is the Petrobras America Cascade FPSO in the Walker Ridge 249 field in 2,600 meters of water.

Main Offshore Fields

Notable offshore fields include:

- the North Sea
- the Gulf of Mexico (offshore Texas, Louisiana, Mississippi, and Alabama)
- California (in the Los Angeles Basin and Santa Barbara Channel, part of the Ventura Basin)
- the Caspian Sea (notably some major fields offshore Azerbaijan)
- the Campos and Santos Basins off the coasts of Brazil
- Newfoundland and Nova Scotia (Atlantic Canada)
- several fields off West Africa most notably west of Nigeria and Angola
- offshore fields in South East Asia and Sakhalin, Russia
- major offshore oil fields are located in the Persian Gulf such as Safaniya, Manifa and Mar-

jan which belong to Saudi Arabia and are developed by Saudi Aramco.

- fields in India (Mumbai High, K G Basin-East Coast Of India, Tapti Field Gujrat, India)

Challenges

Offshore oil and gas production is more challenging than land-based installations due to the remote and harsher environment. Much of the innovation in the offshore petroleum sector concerns overcoming these challenges, including the need to provide very large production facilities. Production and drilling facilities may be very large and a large investment, such as the Troll A platform standing on a depth of 300 meters.

Another type of offshore platform may float with a mooring system to maintain it on location. While a floating system may be lower cost in deeper waters than a fixed platform, the dynamic nature of the platforms introduces many challenges for the drilling and production facilities.

The ocean can add several billion meters or more to the fluid column. The addition increases the equivalent circulating density and downhole pressures in drilling wells, as well as the energy needed to lift produced fluids for separation on the platform.

The trend today is to conduct more of the production operations subsea, by separating water from oil and re-injecting it rather than pumping it up to a platform, or by flowing to onshore, with no installations visible above the sea. Subsea installations help to exploit resources at progressively deeper waters—locations which had been inaccessible—and overcome challenges posed by sea ice such as in the Barents Sea. One such challenge in shallower environments is seabed gouging by drifting ice features (means of protecting offshore installations against ice action includes burial in the seabed).

Offshore manned facilities also present logistics and human resources challenges. An offshore oil platform is a small community in itself with cafeteria, sleeping quarters, management and other support functions. In the North Sea, staff members are transported by helicopter for a two-week shift. They usually receive higher salary than onshore workers do. Supplies and waste are transported by ship, and the supply deliveries need to be carefully planned because storage space on the platform is limited. Today, much effort goes into relocating as many of the personnel as possible onshore, where management and technical experts are in touch with the platform by video conferencing. An onshore job is also more attractive for the aging workforce in the petroleum industry, at least in the western world. These efforts among others are contained in the established term integrated operations. The increased use of subsea facilities helps achieve the objective of keeping more workers onshore. Subsea facilities are also easier to expand, with new separators or different modules for different oil types, and are not limited by the fixed floor space of an above-water installation.

Effects on the Environment

Offshore oil production involves environmental risks, most notably oil spills from oil tankers or pipelines transporting oil from the platform to onshore facilities, and from leaks and accidents on the platform. Produced water is also generated, which is water brought to the surface along with the oil and gas; it is usually highly saline and may include dissolved or unseparated hydrocarbons.

References

- Fjaer, E. (2008). "Mechanics of hydraulic fracturing". Petroleum related rock mechanics. Developments in petroleum science (2nd ed.). Elsevier. p. 369. ISBN 978-0-444-50260-5. Retrieved 2012-05-14.

- Price, N. J.; Cosgrove, J. W. (1990). Analysis of geological structures. Cambridge University Press. pp. 30–33. ISBN 978-0-521-31958-4. Retrieved 5 November 2011.

- Manthei, G.; Eisenblätter, J.; Kamlot, P. (2003). "Stress measurement in salt mines using a special hydraulic fracturing borehole tool". In Natau, Fecker & Pimentel. Geotechnical Measurements and Modelling (PDF). pp. 355–360. ISBN 90-5809-603-3. Retrieved 6 March 2012.

- Gill, R. (2010). Igneous rocks and processes: a practical guide. John Wiley and Sons. p. 102. ISBN 978-1-4443-3065-6. Retrieved 5 November 2011.

- Mader, Detlef (1989). Hydraulic Proppant Fracturing and Gravel Packing. Elsevier. pp. 173–174; 202. ISBN 9780444873521.

- Gold, Russell (2014). The Boom: How Fracking Ignited the American Energy Revolution and Changed the World. New York: Simon & Schuster. pp. 115–121. ISBN 978-1-4516-9228-0.

- Chilingar, George V.; Robertson, John O.; Kumar, Sanjay (1989). Surface Operations in Petroleum Production. 2. Elsevier. pp. 143–152. ISBN 9780444426772.

- Brown, Edwin Thomas (2007) [2003]. Block Caving Geomechanics (2nd ed.). Indooroopilly, Queensland: Julius Kruttschnitt Mineral Research Centre, UQ. ISBN 978-0-9803622-0-6. Retrieved 2012-05-14.

- Zukerman, Gregory (2013-11-06). "Breakthrough: The Accidental Discovery That Revolutionized American Energy". The Atlantis. Retrieved 2016-09-18.

- Brady, Jeff (18 December 2014). "Citing Health, Environment Concerns, New York Moves To Ban Fracking". NPR. Retrieved 25 January 2015.

- V. J. Brown (February 2014). "Radionuclides in Fracking Wastewater: Managing a Toxic Blend". Environmental Health Perspectives. p. A50. Retrieved 27 May 2015.

- Cipolla, Craig (2010). "Hydraulic Fracture Monitoring to Reservoir Simulation: Maximizing Value". SPE Annual Technical Conference and Exhibition. Retrieved 1 January 2014.

- "EU Commission minimum principles for the exploration and production of hydrocarbons (such as shale gas) using high-volume hydraulic fracturing". EUR LEX. Retrieved Nov 2014.

- Negro, Sorrell E. (February 2012). "Fracking Wars: Federal, State, and Local Conflicts over the Regulation of Natural Gas Activities" (PDF). Zoning and Planning Law Report. Thomson Reuters. 35 (2): 1–14. Retrieved 2014-05-01.

- Hweshe, Francis (17 September 2012). "South Africa: International Groups Rally Against Fracking, TKAG Claims". West Cape News. Retrieved 11 February 2014.

- Nicola, Stefan; Andersen, Tino (26 February 2013). "Germany agrees on regulations to allow fracking for shale gas". Bloomberg. Retrieved 1 May 2014.

- "Commission recommendation on minimum principles for the exploration and production of hydrocarbons (such as shale gas) using high-volume hydraulic fracturing (2014/70/EU)". Official Journal of the European Union. 22 January 2014. Retrieved 13 March 2014.

6

Issues and Concerns of Petroleum Exploration

The environmental impact of petroleum exploration is often adverse and affects land and water use, causes noise pollution, water contamination by oil spills and other related problems. This chapter explores the negative impact petroleum exploration has and provides a comprehensive analysis of each detrimental effect. The chapter also provides insightful information on the Hubbert peak theory, Hirsch report and mitigation of peak oil.

Environmental Impact of the Petroleum Industry

The environmental impact of petroleum is often negative because it is toxic to almost all forms of life and its extraction fuels climate change. Petroleum, commonly referred to as oil, is closely linked to virtually all aspects of present society, especially for transportation and heating for both homes and for commercial and industrial activities.

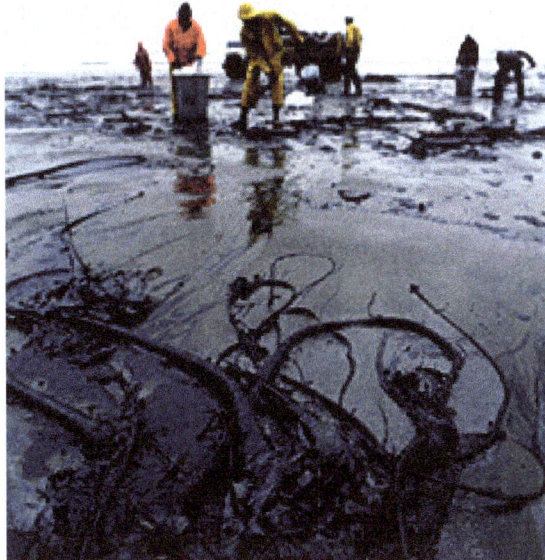

A beach after an oil spill.

Issues

Toxicity

Crude oil is a mixture of many different kinds of organic compounds, many of which are highly toxic and cancer causing (carcinogenic). Oil is "acutely lethal" to fish - that is, it kills fish quickly, at a concentration of 4000 parts per million (ppm) (0.4%). Crude oil and petroleum distillates cause birth defects.

Petroleum distillates can create a sheen on the surface of water as a thin layer creating an optical phenomenon called interphase.

Benzene is present in both crude oil and gasoline and is known to cause leukaemia in humans. The compound is also known to lower the white blood cell count in humans, which would leave people exposed to it more susceptible to infections. "Studies have linked benzene exposure in the mere parts per billion (ppb) range to terminal leukemia, Hodgkin's lymphoma, and other blood and immune system diseases within 5-15 years of exposure."

Exhaust

Petroleum diesel exhaust from a truck

When oil or petroleum distillates are burned usually the combustion is not complete. This means that incompletely burned compounds are created in addition to just water and carbon dioxide. The other compounds are often toxic to life. Examples are carbon monoxide and methanol. Also, fine particulates of soot blacken humans' and other animals' lungs and cause heart problems or death. Soot is cancer causing (carcinogenic).

Acid Rain

Trees killed by acid rain, an unwanted side effect of burning petroleum

High temperatures created by the combustion of petroleum cause nitrogen gas in the surrounding air to oxidize, creating nitrous oxides. Nitrous oxides, along with sulfur dioxide from the sulfur in the oil, combine with water in the atmosphere to create acid rain. Acid rain causes many problems such as dead trees and acidified lakes with dead fish. Coral reefs in the world's oceans are killed by acidic water caused by acid rain.

Acid rain leads to increased corrosion of machinery and structures (large amounts of capital), and to the slow destruction of archaeological structures like the marble ruins in Rome and Greece.

Climate Change

Humans burning large amounts of petroleum create large amounts of CO_2 (carbon dioxide) gas that traps heat in the Earth's atmosphere.

Oil Spills

An oil spill is the release of a liquid petroleum hydrocarbon into the environment, especially marine areas, due to human activity, and is a form of pollution. The term is usually applied to marine oil spills, where oil is released into the ocean or coastal waters, but spills may also occur on land. Oil spills may be due to releases of crude oil from tankers, pipelines, railcars, offshore platforms, drilling rigs and wells, as well as spills of refined petroleum products (such as gasoline, diesel) and their by-products, heavier fuels used by large ships such as bunker fuel, or the spill of any oily refuse or waste oil.

A bird covered in oil from the Black Sea oil spill.

Major oil spills include the Kuwaiti oil fires, Kuwaiti oil lakes, Lakeview Gusher, Gulf War oil spill, and the Deepwater Horizon oil spill. Spilt oil penetrates into the structure of the plumage of birds and the fur of mammals, reducing its insulating ability, and making them more vulnerable to temperature fluctuations and much less buoyant in the water. Cleanup and recovery from an oil spill is difficult and depends upon many factors, including the type of oil spilled, the temperature of the water (affecting evaporation and biodegradation), and the types of shorelines and beaches involved. Spills may take weeks, months or even years to clean up.

Volatile Organic Compounds

Volatile organic compounds (VOCs) are gases or vapours emitted by various solids and liquids, many of which have short- and long-term adverse effects on human health and the environment. VOCs from petroleum are toxic and foul the air, and some like benzene are extremely toxic, carcinogenic and cause DNA damage. Benzene often makes up about 1% of crude oil and gasoline. Benzene is present in automobile exhaust. More important for vapors from spills of diesel and crude oil are aliphatic, volatile compounds. Although "less toxic" than compounds like benzene, their overwhelming abundance can still cause health concerns even when benzene levels in the air are relatively low. The compounds are sometimes collectively measured as "Total Petroleum Hydrocarbons" or "TPH." Petroleum hydrocarbons such as gasoline, diesel, or jet fuel intruding into indoor spaces from underground storage tanks or brownfields threaten safety (e.g., explosive potential) and causes adverse health effects from inhalation.

Waste Oil

Waste oil is used oil containing not only breakdown products but also impurities from use. Some examples of waste oil are used oils such as hydraulic oil, transmission oil, brake fluids, motor oil, crankcase oil, gear box oil and synthetic oil. Many of the same problems associated with natural petroleum exist with waste oil. When waste oil from vehicles drips out engines over streets and roads, the oil travels into the water table bringing with it such toxins as benzene. This poisons both soil and drinking water. Runoff from storms carries waste oil into rivers and oceans, poisoning them as well.

Waste oil in the form of motor oil

Mitigation

Conservation and Phasing Out

- Creating laws to completely phase out the use of petroleum (Sweden's 15-year plan)
- Making use of petroleum more efficiently via better technology

Substitution of Other Energy Sources

- Using "cleaner" energy sources such as natural gas and biodiesel, especially in critical areas like cities where there are people.

Use of Biomass Instead of Petroleum

- It is suggested that cellulose from fibrous plant material, such as hemp, can be used to produce alternatives to many oil-based products.
- Plastics can be created from cellulose instead of from oil.
- Lubricants like motor oil and grease can be made from plants and animal fat.

Safety Measures

- Decreasing the risk of spills
- False floors at gasoline stations to catch gasoline and oil drips from making it into the water table
- Double-hulled tanker ships

Environmental Impact of Hydraulic Fracturing

Hydraulic fracturing
Shale gas drilling rig near Alvarado, Texas
By country
• Canada • New Zealand • South Africa • United Kingdom • United States
Environmental impact
• Additives • United States
Regulation
• United States exemptions
Technology
• Proppants • Uses of radioactivity
Politics

- 2012-14 Romanian protests
- Anti-fracking movement
- *FrackNation*
- Frack Off
- *Gasland*

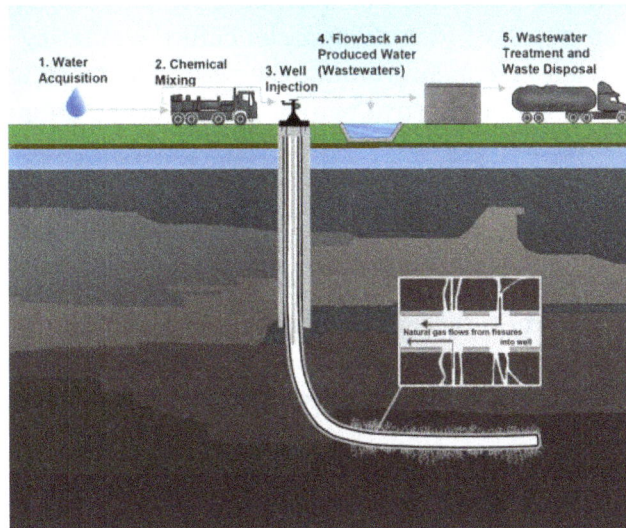

Illustration of hydraulic fracturing and related activities

The environmental impact of hydraulic fracturing affects land use and water consumption, methane emissions, air emissions, water contamination, noise pollution, and health. Water and air pollution are the biggest risks to human health from hydraulic fracturing. Research is underway to determine if human health has been affected, and rigorous adherence to regulation and safety procedures is required to avoid harm. Noise from hydraulic fracturing and associated transport can also affect residents and local wildlife.

Hydraulic fracturing fluids include proppants and other substances, which may include toxic chemicals. In the United States, such additives may be treated as trade secrets by companies who use them. Lack of knowledge about specific chemicals has complicated efforts to develop risk management policies and to study health effects. In other jurisdictions, such as the United Kingdom, these chemicals must be made public and their applications are required to be nonhazardous.

Water usage by hydraulic fracturing can be a problem in areas that experience water shortage. Surface water may be contaminated through spillage and improperly built and maintained waste pits, in jurisdictions where these are permitted. Further, ground water can be contaminated if fluid is able to escape during fracking. Produced water, the water that returns to the surface after fracking, is managed by underground injection, municipal and commercial wastewater treatment, and reuse in future wells. There is potential for methane to leak into ground water and the air, though escape of a methane is a bigger problem in older wells than in those built under more recent legislation.

Hydraulic fracturing causes induced seismicity called microseismic events or microearthquakes. The magnitude of these events is too small to be detected at the surface, being of magnitude M-3 to

M-1 usually. However, fluid disposal wells (which are often used in the USA to dispose of polluted waste from several industries) have been responsible for earthquakes up to 5.6M in Oklahoma and other states.

Governments worldwide are developing regulatory frameworks to assess and manage environmental and associated health risks, working under pressure from industry on the one hand, and from anti-fracking groups on the other. In some countries like France a precautionary approach has been favored and hydraulic fracturing has been banned. Some countries such as the United States have adopted the approach of identifying risks before regulating. The United Kingdom's regulatory framework is based on conclusion that the risks associated with hydraulic fracturing are manageable if carried out under effective regulation and if operational best practices are implemented.

Air Emissions

A report for the European Union on the potential risks was produced in 2012. Potential risks are "methane emissions from the wells, diesel fumes and other hazardous pollutants, ozone precursors or odours from hydraulic fracturing equipment, such as compressors, pumps, and valves". Also gases and hydraulic fracturing fluids dissolved in flowback water pose air emissions risks.

"In the UK, all oil and gas operators must minimise the release of gases as a condition of their licence from the Department of Energy and Climate Change (DECC). Natural gas may only be vented for safety reasons."

Also transportation of necessary water volume for hydraulic fracturing, if done by trucks, can cause emissions. Piped water supplies can reduce the number of truck movements necessary.

A report from the Pennsylvania Dept of Environmental Protection indicated that there is little potential for radiation exposure from oil and gas operations.

Climate Change

Whether natural gas produced by hydraulic fracturing causes higher well-to-burner emissions than gas produced from conventional wells is a matter of contention. Some studies have found that hydraulic fracturing has higher emissions due to methane released during completing wells as some gas returns to the surface, together with the fracturing fluids. Depending on their treatment, the well-to-burner emissions are 3.5%–12% higher than fore conventional gas.

A debate has arisen particularly around a study by professor Robert W. Howarth finding shale gas significantly worse for global warming than oil or coal. Other researchers have criticized Howarth's analysis, including Cathles *et al.,* whose estimates were substantially lower." A 2012 industry funded report co-authored by researchers at the United States Department of Energy's National Renewable Energy Laboratory found emissions from shale gas, when burned for electricity, were "very similar" to those from so-called "conventional well" natural gas, and less than half the emissions of coal.

Several studies which have estimated lifecycle methane leakage from shale gas development and production have found a wide range of leakage rates, from less than 1% of total production to 10%.

According to the Environmental Protection Agency's Greenhouse Gas Inventory a methane leakage rate is about 1.4%. The American Gas Association, an industry trade group, calculated a 1.2% leakage rate. The most comprehensive study of methane leakage from shale gas to date, initiated by the Environmental Defense Fund and released in the Proceedings of the National Academy of Sciences on September 16, 2013, finds that fugitive emissions in key stages of the natural gas production process are significantly lower than estimates in the EPA's national emissions inventory. The study reports direct measurements from 190 onshore natural gas sites, all hydraulically fractured, across the country and estimates a leakage rate of 0.42% for gas production.

Water Consumption

Massive hydraulic fracturing typical of shale wells uses between 1.2 and 3.5 million US gallons (4,500 and 13,200 m^3) of water per well, with large projects using up to 5 million US gallons (19,000 m^3). Additional water is used when wells are refractured. An average well requires 3 to 8 million US gallons (11,000 to 30,000 m^3) of water over its lifetime. According to the Oxford Institute for Energy Studies, greater volumes of fracturing fluids are required in Europe, where the shale depths average 1.5 times greater than in the U.S. Whilst the published amounts may seem large, they are small in comparison with the overall water usage in most areas. A study in Texas, which is a water shortage area, indicates "Water use for shale gas is <1% of statewide water withdrawals; however, local impacts vary with water availability and competing demands."

A report by the Royal Society and the Royal Academy of Engineering shows the usage expected for hydraulic fracturing a well is approximately the amount needed to run a 1,000 MW coal-fired power plant for 12 hours. A 2011 report from the Tyndall Centre estimates that to support a 9 billion cubic metres per annum (320×10^9 cu ft/a) gas production industry, between 1.25 to 1.65 million cubic metres (44×10^6 to 58×10^6 cu ft) would be needed annually, which amounts to 0.01% of the total water abstraction nationally.

Concern has been raised over the increasing quantities of water for hydraulic fracturing in areas that experience water stress. Use of water for hydraulic fracturing can divert water from stream flow, water supplies for municipalities and industries such as power generation, as well as recreation and aquatic life. The large volumes of water required for most common hydraulic fracturing methods have raised concerns for arid regions, such as the Karoo in South Africa, and in droughtprone Texas, in North America. It may also require water overland piping from distant sources.

A 2014 life cycle analysis of natural gas electricity by the National Renewable Energy Laboratory concluded that electricity generated by natural gas from massive hydraulically fractured wells consumed between 249 gallons per megawatt-hour (gal/MWhr) (Marcellus trend) and 272 gal/MWhr (Barnett Shale). The water consumption for the gas from massive hydraulic fractured wells was from 52 to 75 gal/MWhr greater (26 percent to 38 percent greater) than the 197 gal/MWhr consumed for electricity from conventional onshore natural gas.

Some producers have developed hydraulic fracturing techniques that could reduce the need for water. Using carbon dioxide, liquid propane or other gases instead of water have been proposed to reduce water consumption. After it is used, the propane returns to its gaseous state and can be collected and reused. In addition to water savings, gas fracturing reportedly produces less damage to rock formations that can impede production. Recycled flowback water can be reused in hydraulic fracturing. It lowers the total amount of water used and reduces the need to dispose of wastewater after use. The technique is relatively expensive, however, since the water must be treated before each reuse and it can shorten the life of some types of equipment.

Water Contamination

Injected Fluid

In the United States, hydraulic fracturing fluids include proppants, radionuclide tracers, and other chemicals, many of which are toxic. The type of chemicals used in hydraulic fracturing and their properties vary. While most of them are common and generally harmless, some chemicals are carcinogenic. Out of 2,500 products used as hydraulic fracturing additives in the United States, 652 contained one or more of 29 chemical compounds which are either known or possible human carcinogens, regulated under the Safe Drinking Water Act for their risks to human health, or listed as hazardous air pollutants under the Clean Air Act. Another 2011 study identified 632 chemicals used in United States natural gas operations, of which only 353 are well-described in the scientific literature. The Ground Water Protection Council has launched FracFocus.org, an online voluntary disclosure database for hydraulic fracturing fluids funded by oil and gas trade groups and the Department of Energy.

The European Union regulatory regime requires full disclosure of all additives. According to the EU groundwater directive of 2006, "in order to protect the environment as a whole, and human health in particular, detrimental concentrations of harmful pollutants in groundwater must be avoided, prevented or reduced." In the United Kingdom, only chemicals that are "non hazardous in their application" are licensed by the Environment Agency.

Some of the water used in hydraulic fracturing is recovered at the surface as flowback or later production brine. The water left in place is called residual treatment water. According to Engelder and Cathles, this residual treatment water becomes permanently sequestered in the shale and cannot seep into and contaminate ground water.

Flowback

Less than half of injected water is recovered as flowback or later production brine, and in many cases recovery is <30%. As the fracturing fluid flows back through the well, it consists of spent fluids and may contain dissolved constituents such as minerals and brine waters. In some cases, depending on the geology of formation, it may contain uranium, radium, radon and thorium. Estimates of the amount of injected fluid returning to the surface range from 15-20% to 30–70%.

Approaches to managing these fluids, commonly known as produced water, include underground injection, municipal and commercial wastewater treatment and discharge, self-contained systems at well sites or fields, and recycling to fracture future wells. The vacuum multi-effect membrane distillation system as a more effective treatment system has been proposed for treatment of flowback. However, the quantity of waste water needing treatment and the improper configuration of sewage plants have become an issue in some regions of the United States. Part of the wastewater from hydraulic fracturing operations is processed there by public sewage treatment plants, which are not equipped to remove radioactive material and are not required to test for it.

Surface Spills

Surface spills related to the hydraulic fracturing occur mainly because of equipment failure or engineering misjudgments.

Volatile chemicals held in waste water evaporation ponds can to evaporate into the atmosphere, or overflow. The runoff can also end up in groundwater systems. Groundwater may become contaminated by trucks carrying hydraulic fracturing chemicals and wastewater if they are involved in accidents on the way to hydraulic fracturing sites or disposal destinations.

In the evolving European Union legislation, it is required that "Member States should ensure that the installation is constructed in a way that prevents possible surface leaks and spills to soil, water or air." Evaporation and open ponds are not permitted. Regulations call for all pollution pathways to be identified and mitigated. The use of chemical proof drilling pads to contain chemical spills is required. In the UK, total gas security is required, and venting of methane is only permitted in an emergency.

Methane

In September 2014, a study from the US 'Proceedings of the National Academy of Sciences' released a report that indicated that methane contamination can be correlated to distance from a well in wells that were known to leak. This however was not caused by the hydraulic fracturing process, but by poor cementation of casings.

Groundwater methane contamination has adverse effect on water quality and in extreme cases may lead to potential explosion. A scientific study conducted by researchers of Duke University found high correlations of gas well drilling activities, including hydraulic fracturing, and methane pollution of the drinking water. According to the 2011 study of the MIT Energy Initiative, "there is evidence of natural gas (methane) migration into freshwater zones in some areas, most likely as a result of substandard well completion practices i.e. poor quality cementing job or bad casing, by a few operators." A 2013 Duke study suggested that either faulty construction (defective cement seals in the upper part of wells, and faulty steel linings within deeper layers) combined with a peculiarity of local geology may be allowing methane to seep into waters; the latter cause may also release injected fluids to the aquifer. Abandoned gas and oil wells also provide conduits to the surface in areas like Pennsylvania, where these are common.

Some drinking water aquifers naturally contain methane, and drawing down the water level in the aquifer may cause an increase of methane in the drinking water, unrelated to oil or gas drilling. Tests can distinguish between the biogenic methane created by bacteria at shallow depths, and the thermogenic methane, which forms under conditions of high pressure and temperature deeper underground. Most oil and gas development produces the deeper-sourced thermogenic methane. Although methane that occurs naturally in shallow aquifers is usually biogenic, some drinking-water aquifers contain naturally occurring thermogenic methane, or mixed biogenic-thermogenic methane.

A study by Cabot Oil and Gas examined the Duke study using a larger sample size, found that methane concentrations were related to topography, with the highest readings found in low-lying areas, rather than related to distance from gas production areas. Using a more precise isotopic analysis, they showed that the methane found in the water wells came from both the formations where hydraulic fracturing occurred, and from the shallower formations. The Colorado Oil & Gas Conservation Commission investigates complaints from water well owners, and has found some wells to contain biogenic methane unrelated to oil and gas wells, but others that have thermogenic methane due to oil and gas wells with leaking well casing. A review published in February 2012 found no direct evidence that hydraulic fracturing actual injection phase resulted in contamination of ground water, and suggests that reported problems occur due to leaks in its fluid or waste storage apparatus; the review says that methane in water wells in some areas probably comes from natural resources.

Another 2013 review found that hydraulic fracturing technologies are not free from risk of contaminating groundwater, and described the controversy over whether the methane that has been detected in private groundwater wells near hydraulic fracturing sites has been caused by drilling or by natural processes.

Radionuclides

There are naturally occurring radioactive materials (NORM), for example radium, radon, uranium, and thorium, in shale deposits. Brine co-produced and brought to the surface along with the oil and gas sometimes contains naturally occurring radioactive materials; brine from many shale gas wells, contains these radioactive materials. When NORM is concentrated or exposed by human activities, such as hydraulic fracturing, EPA classifies it as TENORM (technologically enhanced naturally occurring radioactive material).

The U.S. Environmental Protection Agency and regulators in North Dakota consider radioactive material in flowback a potential hazard to workers at hydraulic fracturing drilling and waste disposal sites and those living or working nearby if the correct procedures are not followed. A report from the Pennsylvania Department of Environmental Protection indicated that there is little potential for radiation exposure from oil and gas operations.

Land Usage

In the UK, the likely well spacing visualised by the Dec 2013 DECC Strategic Environmental Assessment report indicated that well pad spacings of 5 km were likely in crowded areas, with up to 3 hectares (7.4 acres) per well pad. Each pad could have 24 separate wells. This amounts to 0.16% of land area. A study published in 2015 on the Fayetteville Shale found that a mature gas field impacted about 2% of the land area and substantially increased edge habitat creation. Average land impact per well was 3 hectares (about 7 acres)

Seismicity

Hydraulic fracturing causes induced seismicity called microseismic events or microearthquakes. These microseismic events are often used to map the horizontal and vertical extent of the fracturing. The magnitude of these events is usually too small to be detected at the surface, although the biggest micro-earthquakes may have the magnitude of about -1.5 (M_w).

Induced Seismicity from Hydraulic Fracturing

As of late 2014, there have been three instances of hydraulic fracturing, through induced seismicity, triggering quakes large enough to be felt by people: one each in the United States, Canada, and England. In England, two earthquakes that occurred in April and May 2011 of a magnitude of respectively 1.5 and 2.3 on the Richter scale were felt by local populations. The United Kingdom Department of Energy and Climate Change said the "observed seismicity in April and May 2011 was induced by the hydraulic fracture treatments at Preese Hall", in the North of England.

The National Research Council (part of the National Academy of Sciences) has also observed that hydraulic fracturing, when used in shale gas recovery, does not pose a serious risk of causing earthquakes that can be felt.

Induced Seismicity from Water Disposal Wells

According to the USGS only a small fraction of roughly 30,000 waste fluid disposal wells for oil and gas operations in the United States have induced earthquakes that are large enough to be of concern to the public. Although the magnitudes of these quakes has been small, the USGS says that there is no guarantee that larger quakes will not occur. In addition, the frequency of the quakes has been increasing. In 2009, there were 50 earthquakes greater than magnitude 3.0 in the area spanning Alabama and Montana, and there were 87 quakes in 2010. In 2011 there were 134 earthquakes in the same area, a sixfold increase over 20th century levels. There are also concerns that quakes may damage underground gas, oil, and water lines and wells that were not designed to withstand earthquakes.

Several earthquakes in 2011, including a 4.0 magnitude quake on New Year's Eve that hit Youngstown, Ohio, are likely linked to a disposal of hydraulic fracturing wastewater, according to seismologists at Columbia University. A similar series of small earthquakes occurred in 2012 in Texas. Earthquakes are not common occurrences in either area.

A 2012 US Geological Survey study reported that a "remarkable" increase in the rate of M ≥ 3 earthquakes in the US midcontinent "is currently in progress", having started in 2001 and culminating in a 6-fold increase over 20th century levels in 2011. The overall increase was tied to earthquake increases in a few specific areas: the Raton Basin of southern Colorado (site of coalbed methane activity), and gas-producing areas in central and southern Oklahoma, and central Arkansas. While analysis suggested that the increase is "almost certainly man-made", the USGS noted: "USGS's studies suggest that the actual hydraulic fracturing process is only very rarely the direct cause of felt earthquakes." The increased earthquakes were said to be most likely caused by increased injection of gas-well wastewater into disposal wells. The injection of waste water from oil and gas operations, including from hydraulic fracturing, into saltwater disposal wells may cause bigger low-magnitude tremors, being registered up to 3.3 (M_w).

In 2013, Researchers from Columbia University and the University of Oklahoma demonstrated that in the midwestern United States, some areas with increased human-induced seismicity are susceptible to additional earthquakes triggered by the seismic waves from remote earthquakes. They recommended increased seismic monitoring near fluid injection sites to determine which areas are vulnerable to remote triggering and when injection activity should be ceased.

A British Columbia Oil and Gas Commission investigation concluded that a series of 38 earthquakes (magnitudes ranging from 2.2 to 3.8 on the Richter scale) occurring in the Horn River Basin area between 2009 and 2011 were caused by fluid injection during hydraulic fracturing in proximity to pre-existing faults. The tremors were small enough that only one of them was reported felt by people; there were no reports of injury or property damage.

Noise

Each well pad (in average 10 wells per pad) needs during preparatory and hydraulic fracturing process about 800 to 2,500 days of activity, which may affect residents. In addition, noise is created by transport related to the hydraulic fracturing activities.

The UK Onshore Oil and Gas (UKOOG) is the industry representative body, and it has published a charter that shows how noise concerns will be mitigated, using sound insulation, and heavily silenced rigs where this is needed.

Safety Issues

In July 2013, the United States Federal Railroad Administration listed oil contamination by hydraulic fracturing chemicals as "a possible cause" of corrosion in oil tank cars.

Health Risks

There is worldwide concern over the possible adverse public health implications of hydraulic fracturing activity. A 2013 review on shale gas production in the United States stated, "with increasing numbers of drilling sites, more people are at risk from accidents and exposure to harmful substances used at fractured wells." A 2011 hazard assessment found that most of the chemicals used for hydraulic fracturing and drilling have immediate health effects, and many may have long-term health effects.

In June 2014 Public Health England published a review of the potential public health impacts of exposures to chemical and radioactive pollutants as a result of shale gas extraction in the UK, based on the examination of literature and data from countries where hydraulic fracturing already occurs. The executive summary of the report stated: "An assessment of the currently available evidence indicates that the potential risks to public health from exposure to the emissions associated with shale gas extraction will be low if the operations are properly run and regulated. Most evidence suggests that contamination of groundwater, if it occurs, is most likely to be caused by leakage through the vertical borehole. Contamination of groundwater from the underground hydraulic fracturing process itself (ie the fracturing of the shale) is unlikely. However, surface spills of hydraulic fracturing fluids or wastewater may affect groundwater, and emissions to air also have the potential to impact on health. Where potential risks have been identified in the literature, the reported problems are typically a result of operational failure and a poor regulatory environment."

A 2013 review focusing on Marcellus shale gas hydraulic fracturing and the New York City water supply stated, "Although potential benefits of Marcellus natural gas exploitation are large for transition to a clean energy economy, at present the regulatory framework in New York State is inadequate to prevent potentially irreversible threats to the local environment and New York City water supply. Major investments in state and federal regulatory enforcement will be required to avoid these environmental consequences, and a ban on drilling within the NYC water supply watersheds is appropriate, even if more highly regulated Marcellus gas production is eventually permitted elsewhere in New York State." In 2014, New York State banned hydraulic fracturing entirely, citing health risks.

A 2012 report prepared for the European Union Directorate-General for the Environment identified risks to humans from air pollution and ground water contamination posed by hydraulic fracturing. This led to a series of recommendations in 2014 to mitigate these concerns.

A 2012 guidance for pediatric nurses in the US, said that hydraulic fracturing had a potential negative impact on public health, and that pediatric nurses should be prepared to gather information on such topics so as to advocate for improved community health.

Policy and Science

There are two main approaches to regulation that derive from policy debates about how to manage risk and a corresponding debate about how to assess risk.

The two main schools of regulation are science-based assessment of risk and the taking of measures to prevent harm from those risks through an approach like hazard analysis, and the precautionary principle, where action is taken before risks are well-identified. The relevance and reliability of risk assessments in communities where hydraulic fracturing occurs has also been debated amongst environmental groups, health scientists, and industry leaders. The risks, to some, are overplayed and the current research is insufficient in showing the link between hydraulic fracturing and adverse health effects, while to others the risks are obvious and risk assessment is underfunded.

Different regulatory approaches have thus emerged. In France and Vermont for instance, a precautionary approach has been favored and hydraulic fracturing has been banned based on two principles: the precautionary principle and the prevention principle. Nevertheless, some States such as the U.S. have adopted a risk assessment approach, which had led to many regulatory debates over the issue of hydraulic fracturing and its risks.

In the UK, the regulatory framework is largely being shaped by a report commissioned by the UK Government in 2012, whose purpose was to identify the problems around hydraulic fracturing and to advise the country's regulatory agencies. Jointly published by the Royal Society and the Royal Academy of Engineering, under the chairmanship of Professor Robert Mair, the report features ten recommendations covering issues such as groundwater contamination, well integrity, seismic risk, gas leakages, water management, environmental risks, best practice for risk management, and also includes advice for regulators and research councils. The report was notable for stating that the risks associated with hydraulic fracturing are manageable if carried out under effective regulation and if operational best practices are implemented.

A 2013 review concluded that, in the US, confidentiality requirements dictated by legal investigations have impeded peer-reviewed research into environmental impacts.

Hubbert Peak Theory

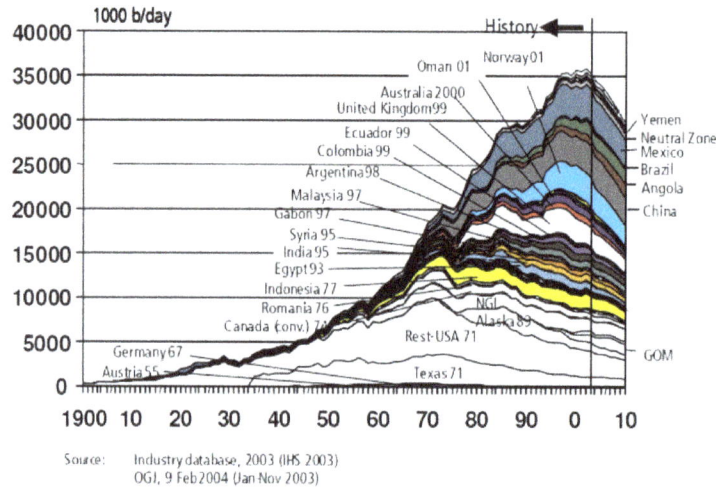

2004 U.S. government predictions for oil production other than in OPEC and the former Soviet Union

The Hubbert peak theory says that for any given geographical area, from an individual oil-producing region to the planet as a whole, the rate of petroleum production tends to follow a bell-shaped curve. It is one of the primary theories on peak oil.

Choosing a particular curve determines a point of maximum production based on discovery rates, production rates and cumulative production. Early in the curve (pre-peak), the production rate increases because of the discovery rate and the addition of infrastructure. Late in the curve (post-peak), production declines because of resource depletion.

The Hubbert peak theory is based on the observation that the amount of oil under the ground in any region is finite, therefore the rate of discovery which initially increases quickly must reach a maximum and decline. In the US, oil extraction followed the discovery curve after a time lag of 32 to 35 years. The theory is named after American geophysicist M. King Hubbert, who created a method of modeling the production curve given an assumed ultimate recovery volume.

Hubbert's Peak

"Hubbert's peak" can refer to the peaking of production of a particular area, which has now been observed for many fields and regions.

Hubbert's Peak was thought to have been achieved in the United States contiguous 48 states (that is, excluding Alaska) in the early 1970s. Oil production peaked at 10,200,000 barrels per day (1,620,000 m³/d) and then declined for several years since. Yet, recent advances in extraction technology, particularly those that led to the extraction of tight oil and oil from shale, have drastically changed the picture. A decline in production followed the 1970s peak, but it has been succeeded by a major increase in production.

Peak oil as a proper noun, or "Hubbert's peak" applied more generally, refers to a predicted event: the peak of the entire planet's oil production. After Peak Oil, according to the Hubbert Peak Theory, the rate of oil production on Earth would enter a terminal decline. On the basis of his theory, in a paper he presented to the American Petroleum Institute in 1956, Hubbert correctly predicted that production of oil from conventional sources would peak in the continental United States around 1965–1970. His prediction of inevitable decline has been incorrect, but the 1970 peak has yet not been surpassed. Hubbert further predicted a worldwide peak at "about half a century" from publication and approximately 12 gigabarrels (GB) a year in magnitude. In a 1976 TV interview Hubbert added that the actions of OPEC might flatten the global production curve but this would only delay the peak for perhaps 10 years. The development of new technologies has provided access to large quantities of unconventional resources, and the boost of production has largely discounted Huppert's prediction. In the future, pressure to limit the use of fossil fuels (and so reduce the release of greenhouse gasses) will curb production, not exhaustion of resources.

Hubbert's Theory

Hubbert Curve

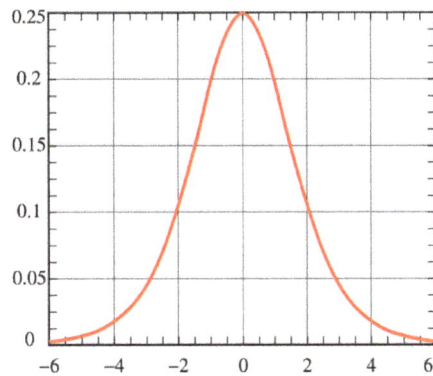

The standard Hubbert curve. For applications, the *x* and *y* scales are replaced by time and production scales.

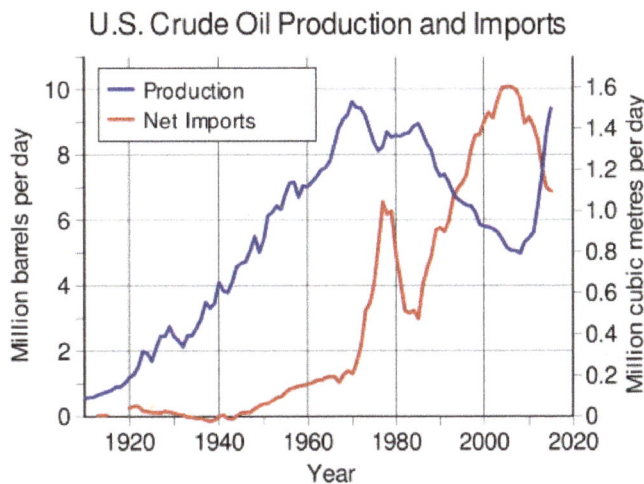

U.S. Oil Production and Imports 1910 to 2012

In 1956, Hubbert proposed that fossil fuel production in a given region over time would follow a roughly bell-shaped curve without giving a precise formula; he later used the Hubbert curve, the derivative of the logistic curve, for estimating future production using past observed discoveries.

Hubbert assumed that after fossil fuel reserves (oil reserves, coal reserves, and natural gas reserves) are discovered, production at first increases approximately exponentially, as more extraction commences and more efficient facilities are installed. At some point, a peak output is reached, and production begins declining until it approximates an exponential decline.

The Hubbert curve satisfies these constraints. Furthermore, it is roughly symmetrical, with the peak of production reached when about half of the fossil fuel that will ultimately be produced has been produced. It also has a single peak.

F Multiple Curves

The sum of multiple Hubbert curves, a technique not developed by Hubbert himself, may be used in order to model more complicated real life scenarios.

Reliability

Crude Oil

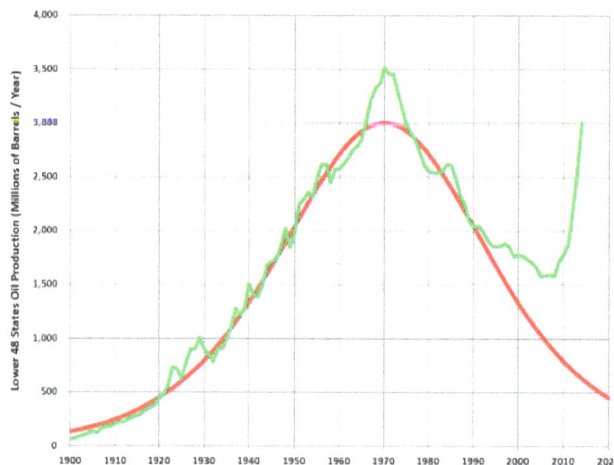

Hubbert's upper-bound prediction for US crude oil production (1956), and actual
lower-48 states production through 2014

Hubbert, in his 1956 paper, presented two scenarios for US crude oil production:

- most likely estimate: a logistic curve with a logistic growth rate equal to 6%, an ultimate resource equal to 150 Giga-barrels (Gb) and a peak in 1965. The size of the ultimate resource was taken from a synthesis of estimates by well-known oil geologists and the US Geological Survey, which Hubbert judged to be the most likely case.

- upper-bound estimate: a logistic curve with a logistic growth rate equal to 6% and ultimate resource equal to 200 Giga-barrels and a peak in 1970.

Hubbert's upper-bound estimate, which he regarded as optimistic, accurately predicted that US oil production would peak in 1970, although the actual peak was 17% higher than Hubbert's curve. Production declined, as Hubbert had predicted, and stayed within 10 percent of Hubbert's predicted value from 1974 through 1994; since then, actual production has been significantly greater than the Hubbert curve. The development of new technologies has provided access to large quantities of unconventional resources, and the boost of production has largely discounted Huppert's prediction. In the future, pressure to limit the use of fossil fuels (and so reduce the release of greenhouse gasses) will curb production, not exhaustion of resources.

Hubbert's 1956 production curves depended on geological estimates of ultimate recoverable oil resources, but he was dissatisfied by the uncertainty this introduced, given the various estimates ranging from 110 billion to 590 billion barrels for the US. Starting in his 1962 publication, he made his calculations, including that of ultimate recovery, based only on mathematical analysis of production rates, proved reserves, and new discoveries, independent of any geological estimates of future discoveries. He concluded that the ultimate recoverable oil resource of the contiguous 48 states was 170 billion barrels, with a production peak in 1966 or 1967. He considered that because his model incorporated past technical advances, that any future advances would occur at the same rate, and were also incorporated. Hubbert continued to defend his calculation of 170 billion barrels in his publications of 1965 and 1967, although by 1967 he had moved the peak forward slightly, to 1968 or 1969.

A post-hoc analysis of peaked oil wells, fields, regions and nations found that Hubbert's model was the "most widely useful" (providing the best fit to the data), though many areas studied had a sharper "peak" than predicted.

A 2007 study of oil depletion by the UK Energy Research Centre pointed out that there is no theoretical and no robust practical reason to assume that oil production will follow a logistic curve. Neither is there any reason to assume that the peak will occur when half the ultimate recoverable resource has been produced; and in fact, empirical evidence appears to contradict this idea. An analysis of a 55 post-peak countries found that the average peak was at 25 percent of the ultimate recovery.

Natural Gas

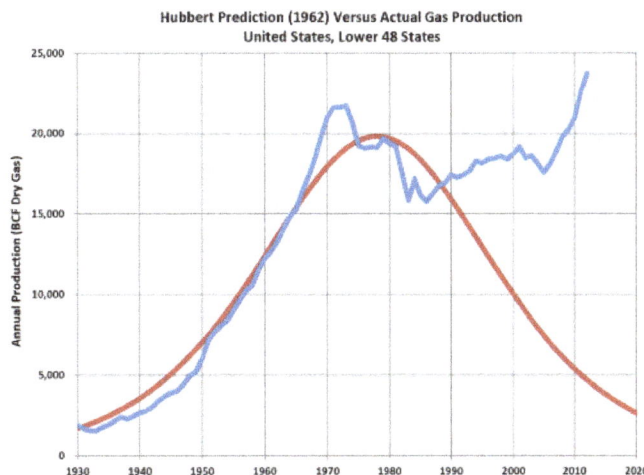

Hubbert's 1962 prediction of US lower 48-state gas production, versus actual production through 2012

Hubbert also predicted that natural gas production would follow a logistic curve similar to that of oil. At right is his gas production curve for the United States, published in 1962.

Economics

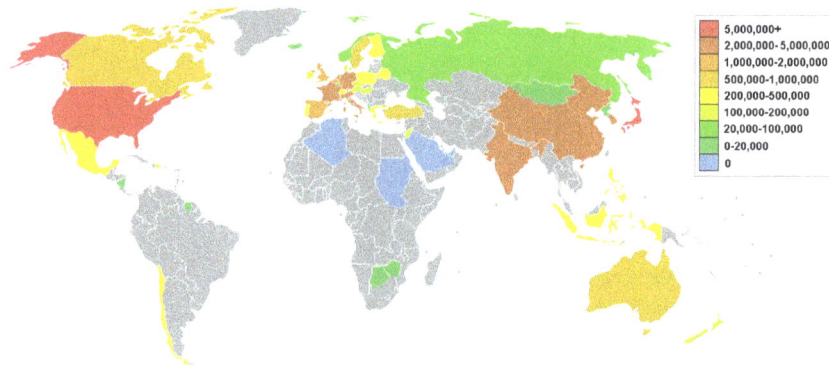

Oil imports by country Pre-2006

Energy Return on Energy Investment

The ratio of energy extracted to the energy expended in the process is often referred to as the Energy Return on Energy Investment (EROI or EROEI). As the EROEI drops to one, or equivalently the Net energy gain falls to zero, the oil production is no longer a net energy source. This happens long before the resource is physically exhausted.

Note that it is important to understand the distinction between a barrel of oil, which is a measure of oil, and a barrel of oil equivalent (BOE), which is a measure of energy. Many sources of energy, such as fission, solar, wind, and coal, are not subject to the same near-term supply restrictions that oil is. Accordingly, even an oil source with an EROEI of 0.5 can be usefully exploited if the energy required to produce that oil comes from a cheap and plentiful energy source. Availability of cheap, but hard to transport, natural gas in some oil fields has led to using natural gas to fuel enhanced oil recovery. Similarly, natural gas in huge amounts is used to power most Athabasca tar sands plants. Cheap natural gas has also led to ethanol fuel produced with a net EROEI of less than 1, although figures in this area are controversial because methods to measure EROEI are in debate.

Advances in technology or experience can lead to greater productivity. The US Energy Information Administration has reported that drilling for shale gas and light tight oil in the United States became much more efficient throughout the period 2007–2014. In terms of oil produced per day of rig drilling time, Bakken wells drilled in January 2014 produced 2.4 times as much oil as those drilled five years earlier, in January 2009. In the Marcellus Gas Trend, wells drilled in January 2014 produced more than nine times as much gas per day of drilling rig time as those drilled five years previously, in January 2009.

Growth-based Economic Models

Insofar as economic growth is driven by oil consumption growth, post-peak societies must adapt. Hubbert believed:

> "Our principal constraints are cultural. During the last two centuries we have known nothing but exponential growth and in parallel we have evolved what amounts to an exponential-growth culture, a culture so heavily dependent upon the continuance of exponential growth for its stability that it is incapable of reckoning with problems of nongrowth."

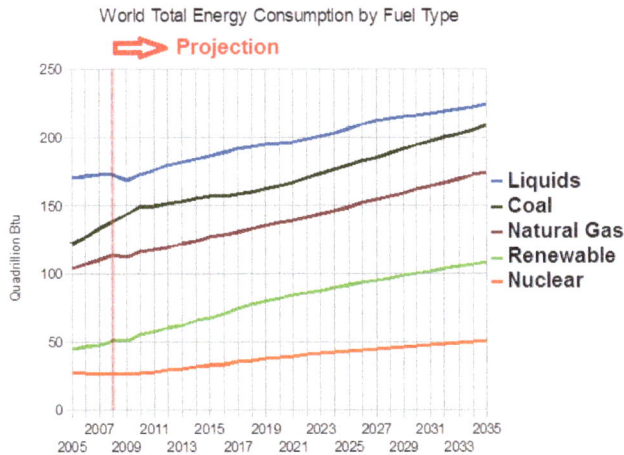

World energy consumption & predictions, 2005–2035. *Source: International Energy Outlook 2011.*

Some economists describe the problem as uneconomic growth or a false economy. At the political right, Fred Ikle has warned about "conservatives addicted to the Utopia of Perpetual Growth". Brief oil interruptions in 1973 and 1979 markedly slowed—but did not stop—the growth of world GDP.

Between 1950 and 1984, as the Green Revolution transformed agriculture around the globe, world grain production increased by 250%. The energy for the Green Revolution was provided by fossil fuels in the form of fertilizers (natural gas), pesticides (oil), and hydrocarbon fueled irrigation.

David Pimentel, professor of ecology and agriculture at Cornell University, and Mario Giampietro, senior researcher at the National Research Institute on Food and Nutrition (INRAN), place in their study *Food, Land, Population and the U.S. Economy* the maximum U.S. population for a sustainable economy at 200 million. To achieve a sustainable economy world population will have to be reduced by two-thirds, says the study. Without population reduction, this study predicts an agricultural crisis beginning in 2020, becoming critical c. 2050. The peaking of global oil along with the decline in regional natural gas production may precipitate this agricultural crisis sooner than generally expected. Dale Allen Pfeiffer claims that coming decades could see spiraling food prices without relief and massive starvation on a global level such as never experienced before.

Hubbert Peaks

Although Hubbert peak theory receives most attention in relation to peak oil production, it has also been applied to other natural resources.

Natural Gas

Doug Reynolds predicted in 2005 that the North American peak would occur in 2007. Bentley predicted a world "decline in conventional gas production from about 2020".

Coal

Although observers believe that peak coal is significantly further out than peak oil, Hubbert studied the specific example of anthracite in the USA, a high grade coal, whose production peaked in the 1920s. Hubbert found that anthracite matches a curve closely. Hubbert had recoverable coal reserves worldwide at 2.500×10^{12} metric tons and peaking around 2150 (depending on usage).

More recent estimates suggest an earlier peak. *Coal: Resources and Future Production* (PDF 630KB), published on April 5, 2007 by the Energy Watch Group (EWG), which reports to the German Parliament, found that global coal production could peak in as few as 15 years. Reporting on this, Richard Heinberg also notes that the date of peak annual energetic extraction from coal is likely to come earlier than the date of peak in quantity of coal (tons per year) extracted as the most energy-dense types of coal have been mined most extensively. A second study, *The Future of Coal* by B. Kavalov and S. D. Peteves of the Institute for Energy (IFE), prepared for European Commission Joint Research Centre, reaches similar conclusions and states that ""coal might not be so abundant, widely available and reliable as an energy source in the future".

Work by David Rutledge of Caltech predicts that the total of world coal production will amount to only about 450 gigatonnes. This implies that coal is running out faster than usually assumed.

Fissionable Materials

In a paper in 1956, after a review of US fissionable reserves, Hubbert notes of nuclear power:

> "There is promise, however, provided mankind can solve its international problems and not destroy itself with nuclear weapons, and provided world population (which is now expanding at such a rate as to double in less than a century) can somehow be brought under control, that we may at last have found an energy supply adequate for our needs for at least the next few centuries of the "foreseeable future.""

Technologies such as the thorium fuel cycle, reprocessing and fast breeders can, in theory, considerably extend the life of uranium reserves. Roscoe Bartlett claimed in 2006 that

> "Our current throwaway nuclear cycle uses up the world reserve of low-cost uranium in about 20 years."

Caltech physics professor David Goodstein stated in 2004 that

> "... you would have to build 10,000 of the largest power plants that are feasible by engineering standards in order to replace the 10 terawatts of fossil fuel we're burning today ... that's a staggering amount and if you did that, the known reserves of uranium would last for 10 to 20 years at that burn rate. So, it's at best a bridging technology ... You can use the rest of the uranium to breed plutonium 239 then we'd have at least 100 times as much fuel to use. But that means you're making plutonium, which is an extremely dangerous thing to do in the dangerous world that we live in."

Helium

Almost all helium on Earth is a result of radioactive decay of uranium and thorium. Helium is extracted by fractional distillation from natural gas, which contains up to 7% helium. The world's largest helium-rich natural gas fields are found in the United States, especially in the Hugoton and nearby gas fields in Kansas, Oklahoma, and Texas. The extracted helium is stored underground in the National Helium Reserve near Amarillo, Texas, the self-proclaimed "Helium Capital of the World". Helium production is expected to decline along with natural gas production in these areas.

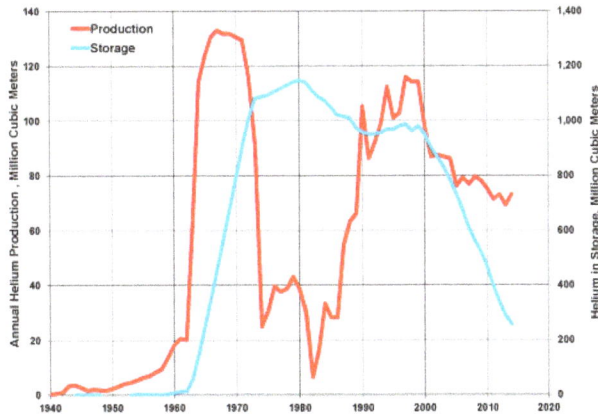

Helium production and storage in the United States, 1940–2014 (data from USGS)

Helium, which is the second-lightest chemical element, will rise to the upper layers of Earth's atmosphere, where it can forever break free from Earth's gravitational attraction. Approximately 1,600 tons of helium are lost per year as a result of atmospheric escape mechanisms.

Transition Metals

Hubbert applied his theory to "rock containing an abnormally high concentration of a given metal" and reasoned that the peak production for metals such as copper, tin, lead, zinc and others would occur in the time frame of decades and iron in the time frame of two centuries like coal. The price of copper rose 500% between 2003 and 2007 and was attributed by some to peak copper. Copper prices later fell, along with many other commodities and stock prices, as demand shrank from fear of a global recession. Lithium availability is a concern for a fleet of Li-ion battery using cars but a paper published in 1996 estimated that world reserves are adequate for at least 50 years. A similar prediction for platinum use in fuel cells notes that the metal could be easily recycled.

Precious Metals

In 2009, Aaron Regent president of the Canadian gold giant Barrick Gold said that global output has been falling by roughly one million ounces a year since the start of the decade. The total global mine supply has dropped by 10pc as ore quality erodes, implying that the roaring bull market of the last eight years may have further to run. "There is a strong case to be made that we are already at 'peak gold'," he told The Daily Telegraph at the RBC's annual gold conference in London. "Production peaked around 2000 and it has been in decline ever since, and we forecast that decline to continue. It is increasingly difficult to find ore," he said.

Ore grades have fallen from around 12 grams per tonne in 1950 to nearer 3 grams in the US, Canada, and Australia. South Africa's output has halved since peaking in 1970. Output fell a further 14 percent in South Africa in 2008 as companies were forced to dig ever deeper – at greater cost – to replace depleted reserves.

World mined gold production has peaked four times since 1900: in 1912, 1940, 1971, and 2001, each peak being higher than previous peaks. The latest peak was in 2001, when production reached 2,600 metric tons, then declined for several years. Production started to increase again in 2009, spurred by high gold prices, and achieved record new highs each year in 2012, 2013, and in 2014, when production reached 2,990 tonnes.

Phosphorus

Phosphorus supplies are essential to farming and depletion of reserves is estimated at somewhere from 60 to 130 years. According to a 2008 study, the total reserves of phosphorus are estimated to be approximately 3,200 MT, with a peak production at 28 MT/year in 2034. Individual countries' supplies vary widely; without a recycling initiative America's supply is estimated around 30 years. Phosphorus supplies affect agricultural output which in turn limits alternative fuels such as bio-diesel and ethanol. Its increasing price and scarcity (global price of rock phosphate rose 8-fold in the 2 years to mid 2008) could change global agricultural patterns. Lands, perceived as marginal because of remoteness, but with very high phosphorus content, such as the Gran Chaco may get more agricultural development, while other farming areas, where nutrients are a constraint, may drop below the line of profitability.

Peak Water

Hubbert's original analysis did not apply to renewable resources. However, over-exploitation often results in a Hubbert peak nonetheless. A modified Hubbert curve applies to any resource that can be harvested faster than it can be replaced.

For example, a reserve such as the Ogallala Aquifer can be mined at a rate that far exceeds replenishment. This turns much of the world's underground water and lakes into finite resources with peak usage debates similar to oil. These debates usually center around agriculture and suburban water usage but generation of electricity from nuclear energy or coal and tar sands mining mentioned above is also water resource intensive. The term fossil water is sometimes used to describe aquifers whose water is not being recharged.

Renewable Resources

- Fisheries: At least one researcher has attempted to perform Hubbert linearization (Hubbert curve) on the whaling industry, as well as charting the transparently dependent price of caviar on sturgeon depletion. The Atlantic northwest cod fishery was a renewable resource, but the numbers of fish taken exceeded the fish's rate of recovery. The end of the cod fishery does match the exponential drop of the Hubbert bell curve. Another example is the cod of the North Sea.

- Air/oxygen: Half the world's oxygen is produced by phytoplankton. The numbers of plankton have dropped by 40% since the 1950s.

Criticisms of Peak Oil

Economist Michael Lynch argues that the theory behind the Hubbert curve is too simplistic and relies on an overly Malthusian point of view. Lynch claims that Campbell's predictions for world oil production are strongly biased towards underestimates, and that Campbell has repeatedly pushed back the date.

Leonardo Maugeri, vice president of the Italian energy company Eni, argues that nearly all of peak estimates do not take into account unconventional oil even though the availability of these resources is significant and the costs of extraction and processing, while still very high, are falling because of improved technology. He also notes that the recovery rate from existing world oil fields has increased from about 22% in 1980 to 35% today because of new technology and predicts this trend will continue. The ratio between proven oil reserves and current production has constantly improved, passing from 20 years in 1948 to 35 years in 1972 and reaching about 40 years in 2003. These improvements occurred even with low investment in new exploration and upgrading technology because of the low oil prices during the last 20 years. However, Maugeri feels that encouraging more exploration will require relatively high oil prices.

Edward Luttwak, an economist and historian, claims that unrest in countries such as Russia, Iran and Iraq has led to a massive underestimate of oil reserves. The Association for the Study of Peak Oil and Gas (ASPO) responds by claiming neither Russia nor Iran are troubled by unrest currently, but Iraq is.

Cambridge Energy Research Associates authored a report that is critical of Hubbert-influenced predictions:

> "Despite his valuable contribution, M. King Hubbert's methodology falls down because it does not consider likely resource growth, application of new technology, basic commercial factors, or the impact of geopolitics on production. His approach does not work in all cases-including on the United States itself-and cannot reliably model a global production outlook. Put more simply, the case for the imminent peak is flawed. As it is, production in 2005 in the Lower 48 in the United States was 66 percent higher than Hubbert projected."

CERA does not believe there will be an endless abundance of oil, but instead believes that global production will eventually follow an "undulating plateau" for one or more decades before declining slowly, and that production will reach 40 Mb/d by 2015.

Alfred J. Cavallo, while predicting a conventional oil supply shortage by no later than 2015, does not think Hubbert's peak is the correct theory to apply to world production.

Criticisms of Peak Element Scenarios

Although M. King Hubbert himself made major distinctions between decline in petroleum production versus depletion (or relative lack of it) for elements such as fissionable uranium and thorium, some others have predicted peaks like peak uranium and peak phosphorus soon on the basis of published reserve figures compared to present and future production. According to some economists, though, the amount of proved reserves inventoried at a time may be considered "a poor indicator of the total future supply of a mineral resource."

As some illustrations, tin, copper, iron, lead, and zinc all had both production from 1950 to 2000 and reserves in 2000 much exceed world reserves in 1950, which would be impossible except for how "proved reserves are like an inventory of cars to an auto dealer" at a time, having little relationship to the actual total affordable to extract in the future. In the example of peak phosphorus, additional concentrations exist intermediate between 71,000 Mt of identified reserves (USGS) and the approximately 30,000,000,000 Mt of other phosphorus in Earth's crust, with the average rock being 0.1% phosphorus, so showing decline in human phosphorus production will occur soon would require far more than comparing the former figure to the 190 Mt/year of phosphorus extracted in mines (2011 figure).

Peak Oil

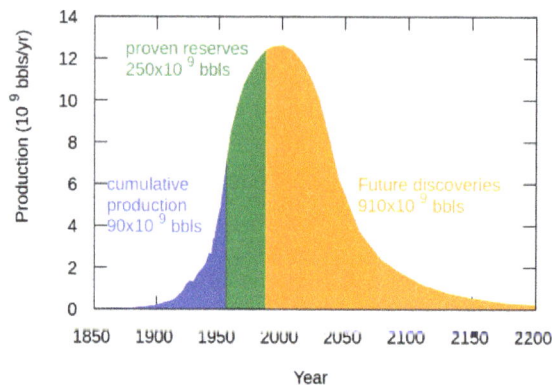

A 1956 world oil production distribution, showing historical data and future production, proposed by M. King Hubbert – it has a peak of 12.5 billion barrels per year in about the year 2000

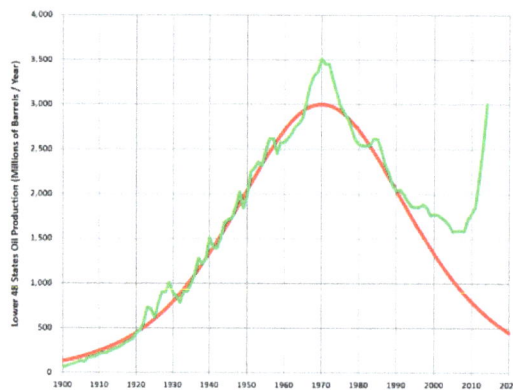

Hubbert's upper-bound prediction for US crude oil production (1956), and actual lower-48 states production through 2014

Peak oil, an event based on M. King Hubbert's theory, is the point in time when the maximum rate of extraction of petroleum is reached, after which it is expected to enter terminal decline. Peak oil theory is based on the observed rise, peak, fall, and depletion of aggregate production rate in oil fields over time. It is often confused with oil depletion; however, peak oil is the point of maximum production, while depletion refers to a period of falling reserves and supply.

Some observers, such as petroleum industry experts Kenneth S. Deffeyes and Matthew Simmons, predict negative global economy implications following a post-peak production decline and subsequent oil price increase because of the high dependence of most modern industrial transport, agricultural, and industrial systems on the low cost and high availability of oil. Predictions vary greatly as to what exactly these negative effects would be.

Oil production forecasts on which predictions of peak oil are based are often made within a range which includes optimistic (higher production) and pessimistic (lower production) scenarios. Optimistic estimations of peak production forecast the global decline will begin after 2020, and assume major investments in alternatives will occur before a crisis, without requiring major changes in the lifestyle of heavily oil-consuming nations. Pessimistic predictions of future oil production made after 2007 stated either that the peak had already occurred, that oil production was on the cusp of the peak, or that it would occur shortly.

Hubbert's original prediction that US peak oil would be in about 1970 seemed accurate for a time, as US average annual production peaked in 1970 at 9.6 million barrels per day. However, the successful application of massive hydraulic fracturing to additional tight reservoirs caused US production to rebound, challenging the inevitability of post-peak decline for the US oil production. In addition, Hubbert's original predictions for world peak oil production proved premature.

Modeling Global Oil Production

The idea that the rate of oil production would peak and irreversibly decline is an old one. In 1919, David White, chief geologist of the United States Geological Survey, wrote of US petroleum: "... the peak of production will soon be passed, possibly within 3 years." In 1953, Eugene Ayers, a researcher for Gulf Oil, projected that if US ultimate recoverable oil reserves were 100 billion barrels, then production in the US would peak no later than 1960. If ultimate recoverable were to be as high as 200 billion barrels, which he warned was wishful thinking, US peak production would come no later than 1970. Likewise for the world, he projected a peak somewhere between 1985 (one trillion barrels ultimate recoverable) and 2000 (two trillion barrels recoverable). Ayers made his projections without a mathematical model. He wrote: "But if the curve is made to look reasonable, it is quite possible to adapt mathematical expressions to it and to determine, in this way, the peak dates corresponding to various ultimate recoverable reserve numbers"

By observing past discoveries and production levels, and predicting future discovery trends, the geoscientist M. King Hubbert used statistical modelling in 1956 to accurately predict that United States oil production would peak between 1965 and 1971. Hubbert used a semi-logistical curved model (sometimes incorrectly compared to a normal distribution). He assumed the production rate of a limited resource would follow a roughly symmetrical distribution. Depending on the limits of exploitability and market pressures, the rise or decline of resource production over time might be sharper or more stable, appear more linear or curved. That model and its variants are now called Hubbert peak theory; they have been used to describe and predict the peak and decline of production from regions, countries, and multinational areas. The same theory has also been applied to other limited-resource production.

In a 2006 analysis of Hubbert theory, it was noted that uncertainty in real world oil production amounts and confusion in definitions increases the uncertainty in general of production predictions. By comparing the fit of various other models, it was found that Hubbert's methods yielded the closest fit over all, but that none of the models were very accurate. In 1956 Hubbert himself recommended using "a family of possible production curves" when predicting a production peak and decline curve.

More recently, the term "peak oil" was popularized by Colin Campbell and Kjell Aleklett in 2002 when they helped form the Association for the Study of Peak Oil and Gas (ASPO). In his publications, Hubbert used the term "peak production rate" and "peak in the rate of discoveries".

Demand

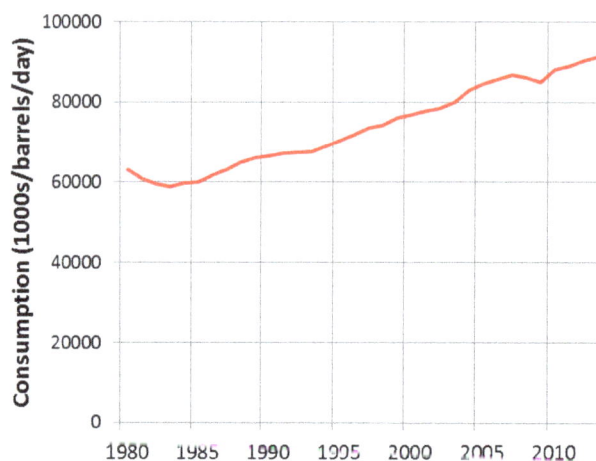

Global consumption of oil 1980–2013 (Energy Information Administration)

The demand side of peak oil over time is concerned with the total quantity of oil that the global market would choose to consume at various possible market prices and how this entire listing of quantities at various prices would evolve over time. Global demand for crude oil grew an average of 1.76% per year from 1994 to 2006, with a high growth of 3.4% in 2003–2004. After reaching a high of 85.6 million barrels (13,610,000 m³) per day in 2007, world consumption decreased in both 2008 and 2009 by a total of 1.8%, despite fuel costs plummeting in 2008. Despite this lull, world quantity-demanded for oil is projected to increase 21% over 2007 levels by 2030 (104 million barrels per day (16.5×10^6 m³/d) from 86 million barrels (13.7×10^6 m³)), or about 0.8% average annual growth, due in large part to increases in demand from the transportation sector. According to projections by the International Energy Agency (IEA) in 2013, growth in global oil demand will be significantly outpaced by growth in production capacity over the next 5 years. Developments in late 2014–2015 have seen an oversupply of global markets leading to a significant drop in the price of oil.

Energy demand is distributed amongst four broad sectors: transportation, residential, commercial, and industrial. In terms of oil use, transportation is the largest sector and the one that has seen the largest growth in demand in recent decades. This growth has largely come from new demand for personal-use vehicles powered by internal combustion engines. This sector also has the highest consumption rates, accounting for approximately 71% of the oil used in the United States in 2013. and 55% of oil use worldwide as documented in the Hirsch report. Transportation is therefore of particular interest to those seeking to mitigate the effects of peak oil.

Although demand growth is highest in the developing world, the United States is the world's largest consumer of petroleum. Between 1995 and 2005, US consumption grew from 17,700,000 barrels per day (2,810,000 m³/d) to 20,700,000 barrels per day (3,290,000 m³/d), a 3,000,000 barrels per day (480,000 m³/d) increase. China, by comparison, increased consumption from 3,400,000 barrels per day (540,000 m³/d) to 7,000,000 barrels per day (1,100,000 m³/d), an increase of 3,600,000 barrels per day (570,000 m³/d), in the same time frame. The Energy Information Administration (EIA) stated that gasoline usage in the United States may have peaked in 2007, in part because of increasing interest in and mandates for use of biofuels and energy efficiency.

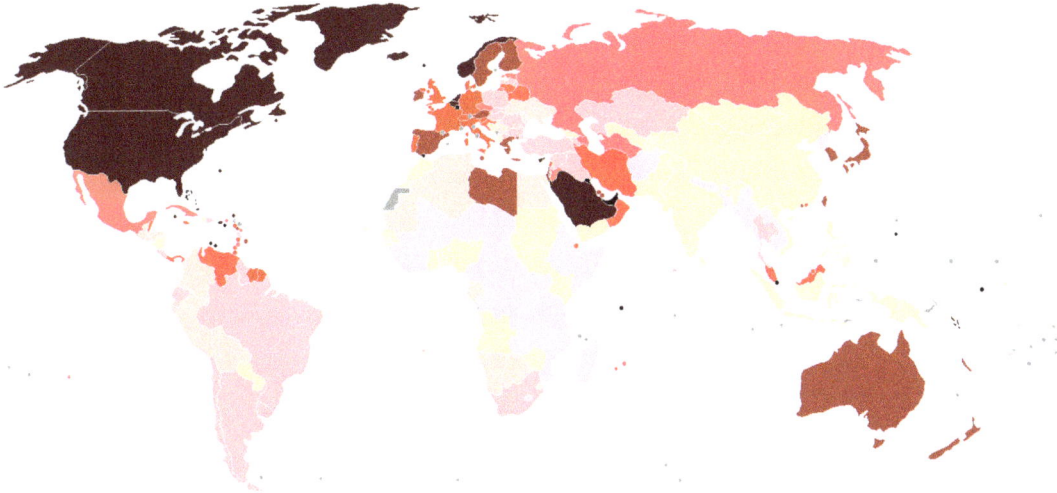

Oil consumption per capita (darker colors represent more consumption, gray represents no data)

As countries develop, industry and higher living standards drive up energy use, oil usage being a major component. Thriving economies, such as China and India, are quickly becoming large oil consumers. For example, China surpassed the United States as the world's largest crude oil importer in 2015. Oil consumption growth is expected to continue; however, not at previous rates, as China's economic growth is predicted to decrease from the high rates of the early part of the 21st century. India's oil imports are expected to more than triple from 2005 levels by 2020, rising to 5 million barrels per day (790×103 m³/d).

Population

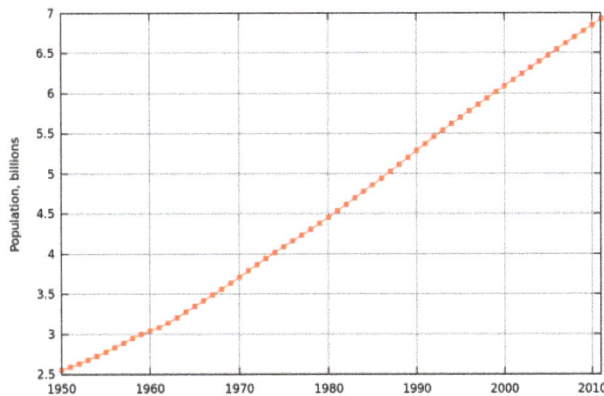

World population

Another significant factor affecting petroleum demand has been human population growth. The United States Census Bureau predicts that world population in 2030 will be almost double that of 1980. Oil production per capita peaked in 1979 at 5.5 barrels/year but then declined to fluctuate around 4.5 barrels/year since. In this regard, the decreasing population growth rate since the 1970s has somewhat ameliorated the per capita decline.

Economic Growth

Some analysts argue that the cost of oil has a profound effect on economic growth due to its pivotal role in the extraction of resources and the processing, manufacturing, and transportation of goods. As the industrial effort to extract new unconventional oil sources increases, this has a compounding negative effect on all sectors of the economy, leading to economic stagnation or even eventual contraction. Such a scenario would result in an inability for national economies to pay high oil prices, leading to declining demand and a price collapse.

Supply

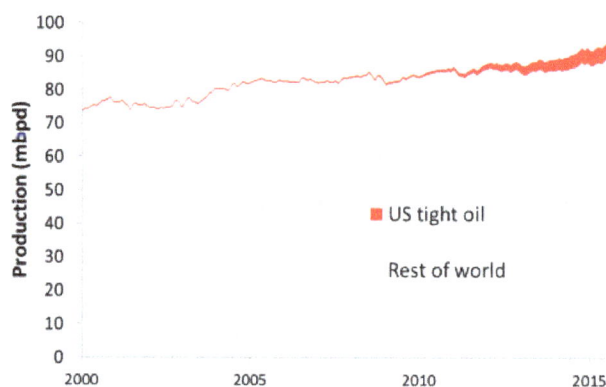

Global liquids production 2000-2015, indicating the component of US tight oil (Energy Information Administration)

Our analysis suggests there are ample physical oil and liquid fuel resources for the foreseeable future. However, the rate at which new supplies can be developed and the break-even prices for those new supplies are changing.

—International Energy Agency

Defining Sources of Oil

Oil may come from conventional or unconventional sources. The terms are not strictly defined, and vary within the literature as definitions based on new technologies tend to change over time. As a result, different oil forecasting studies have included different classes of liquid fuels. Some use the terms "conventional" oil for what is included in the model, and "unconventional" oil for classes excluded.

In 1956, Hubbert confined his peak oil prediction to that crude oil "producible by methods now in use." By 1962, however, his analyses included future improvements in exploration and production. All of Hubbert's analyses of peak oil specifically excluded oil manufactured from oil shale or mined from oil sands. A 2013 study predicting an early peak excluded deepwater oil, tight oil, oil with

API gravity less than 17.5, and oil close to the poles, such as that on the North Slope of Alaska, all of which it defined as non-conventional. Some commonly used definitions for conventional and unconventional oil are detailed below.

Conventional Sources

Conventional oil is extracted on land and offshore using standard techniques, and can be categorized as light, medium, heavy, or extra heavy in grade. The exact definitions of these grades vary depending on the region from which the oil came. Light oil flows naturally to the surface or can be extracted by simply pumping it out of the ground. Heavy refers to oil that has higher density and therefore lower API gravity. It does not flow easily, and its consistency is similar to that of molasses. While some of it can be produced using conventional techniques, recovery rates are better using unconventional methods.

Unconventional Sources

Oil currently considered unconventional is derived from multiple sources.

- Tight oil is extracted from deposits of low-permeability rock, sometimes shale deposits but often other rock types, using hydraulic fracturing, or "fracking." It is often confused with shale oil, which is oil manufactured from the kerogen contained in an oil shale, Production of tight oil has led to a resurgence of US production in recent years. However, tight oil production peaked in 2015 and is not expected to increase again until there is a significant oil price recovery.

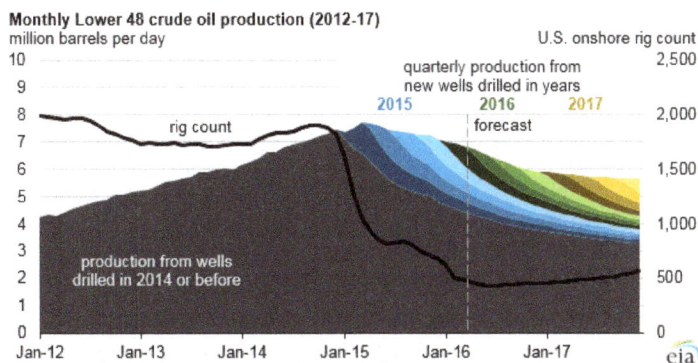

US Lower 48 oil production from 2012 and anticipated decline in production to the end of 2017, with rig count (Energy Information Administration)

- Oil shale is a common term for sedimentary rock such as shale or marl, containing kerogen, a waxy oil precursor that has not yet been transformed into crude oil by the high pressures and temperatures caused by deep burial. The term "oil shale" is somewhat confusing, because what is referred to in the U.S. as "oil shale" is not really oil and the rock it is found in is generally not shale. Since it is close to the surface rather than buried deep in the earth, the shale or marl is typically mined, crushed, and retorted, producing synthetic oil from the kerogen. Its net energy yield is much lower than conventional oil, so much so that estimates of the net energy yield of shale discoveries are considered extremely unreliable.

- Oil sands are unconsolidated sandstone deposits containing large amounts of very viscous crude bitumen or extra-heavy crude oil that can be recovered by surface mining or by in-situ oil wells using steam injection or other techniques. It can be liquefied by upgrading, blending with diluent, or by heating; and then processed by a conventional oil refinery. The recovery process requires advanced technology but is more efficient than that of oil shale. The reason is that, unlike U.S. "oil shale", Canadian oil sands actually contain oil, and the sandstones they are found in are much easier to produce oil from than shale or marl. In the U.S. dialect of English, these formations are often called "tar sands", but the material found in them is not tar but an extra-heavy and viscous form of oil technically known as bitumen.

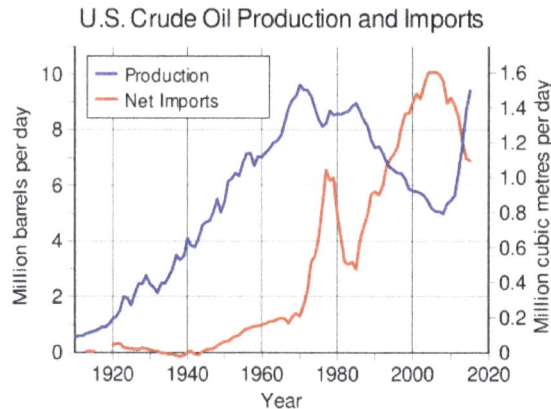

United States crude oil production exceeds imports for the first time since the early 1990s

- Coal liquefaction or gas to liquids product are liquid hydrocarbons that are synthesised from the conversion of coal or natural gas by the Fischer-Tropsch process, Bergius process, or Karrick process. Currently, two companies SASOL and Shell, have synthetic oil technology proven to work on a commercial scale. Sasol's primary business is based on CTL (coal-to-liquid) and GTL (natural gas-to-liquid) technology, producing US$4.40 billion in revenues (FY2009). Shell has used these processes to recycle waste flare gas (usually burnt off at oil wells and refineries) into usable synthetic oil. However, for CTL there may be insufficient coal reserves to supply global needs for both liquid fuels and electric power generation.

- Minor sources include thermal depolymerization, as discussed in a 2003 article in *Discover* magazine, that could be used to manufacture oil indefinitely, out of garbage, sewage, and agricultural waste. The article claimed that the cost of the process was $15 per barrel. A follow-up article in 2006 stated that the cost was actually $80 per barrel, because the feedstock that had previously been considered as hazardous waste now had market value. A 2008 news bulletin published by Los Alamos Laboratory proposed that hydrogen (possibly produced using hot fluid from nuclear reactors to split water into hydrogen and oxygen) in combination with sequestered CO_2 could be used to produce methanol (CH_3OH), which could then be converted into gasoline.

Discoveries

All the easy oil and gas in the world has pretty much been found. Now comes the harder work in finding and producing oil from more challenging environments and work areas.

— William J. Cummings, Exxon-Mobil company spokesman, December 2005

It is pretty clear that there is not much chance of finding any significant quantity of new cheap oil. Any new or unconventional oil is going to be expensive.

— Lord Ron Oxburgh, a former chairman of Shell, October 2008

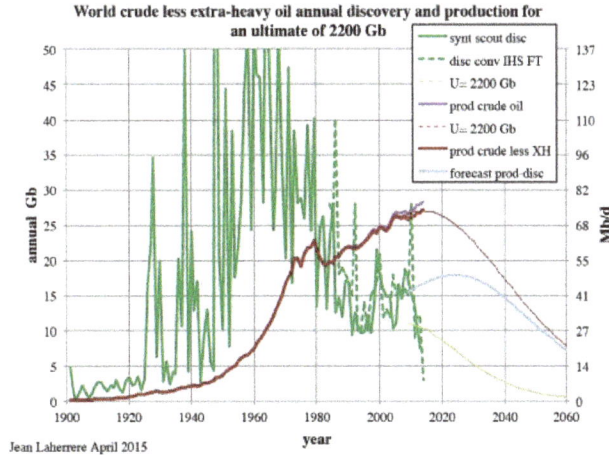

World oil discoveries peaked in the 1960s

The peak of world oilfield discoveries occurred in the 1960s at around 55 billion barrels (8.7×10^9 m³) (Gb)/year. According to the Association for the Study of Peak Oil and Gas (ASPO), the rate of discovery has been falling steadily since. Less than 10 Gb/yr of oil were discovered each year between 2002 and 2007. According to a 2010 Reuters article, the annual rate of discovery of new fields has remained remarkably constant at 15–20 Gb/yr.

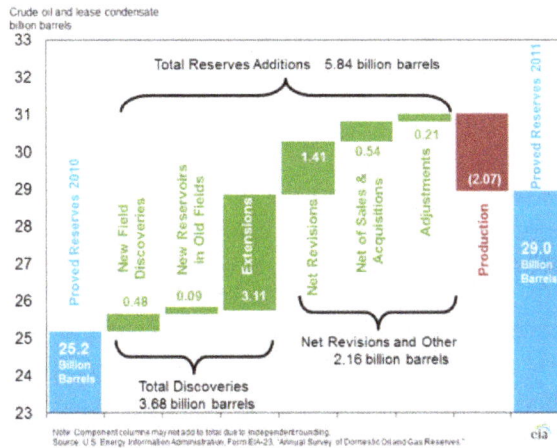

Although US proved oil reserves grew by 3.8 billion barrels in 2011, even after deducting 2.07 billion barrels of production, only 8 percent of the 5.84 billion barrels of the newly booked oil was because of new field discoveries (U.S. EIA)

But despite the fall-off in new field discoveries, and record-high production rates, the reported proved reserves of crude oil remaining in the ground in 2014, which totaled 1,490 billion barrels, not counting Canadian heavy oil sands, were more than quadruple the 1965 proved reserves of 354 billion barrels. A researcher for the U.S. Energy Information Administration has pointed out that after the first wave of discoveries in an area, most oil and natural gas reserve growth comes not from discoveries of new fields, but from extensions and additional gas found within existing fields.

A report by the UK Energy Research Centre noted that "discovery" is often used ambiguously, and explained the seeming contradiction between falling discovery rates since the 1960s and increasing reserves by the phenomenon of reserve growth. The report noted that increased reserves within a field may be discovered or developed by new technology years or decades after the original discovery. But because of the practice of "backdating," any new reserves within a field, even those to be discovered decades after the field discovery, are attributed to the year of initial field discovery, creating an illusion that discovery is not keeping pace with production.

Reserves

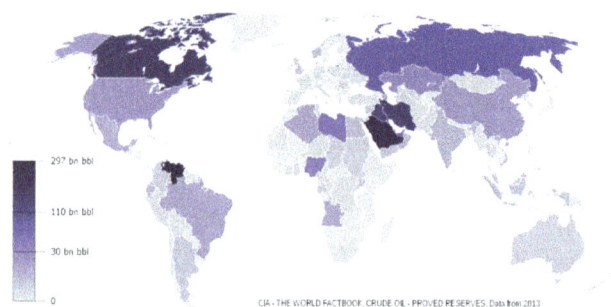

Proven oil reserves, 2013

Total possible conventional crude oil reserves include crude oil with 90–95% certainty of being technically possible to produce (from reservoirs through a wellbore using primary, secondary, improved, enhanced, or tertiary methods); all crude with a 50% probability of being produced in the future; and discovered reserves that have a 5–10% possibility of being produced in the future. These are referred to as 1P/Proven (90–95%), 2P/Probable (50%), and 3P/Possible (5–10%), respectively. This does not include liquids extracted from mined solids or gasses (oil sands, oil shale, gas-to-liquid processes, or coal-to-liquid processes).

Hubbert's 1956 peak projection for the United States depended on geological estimates of ultimate recoverable oil resources, but starting in his 1962 publication, he concluded that ultimate oil recovery was an output of his mathematical analysis, rather than an assumption. He regarded his peak oil calculation as independent of reserve estimates.

Many current 2P calculations predict reserves to be between 1150 and 1350 Gb, but some authors have written that because of misinformation, withheld information, and misleading reserve calculations, 2P reserves are likely nearer to 850–900 Gb. The Energy Watch Group wrote that actual reserves peaked in 1980, when production first surpassed new discoveries, that apparent increases in reserves since then are illusory, and concluded (in 2007): "Probably the world oil production has peaked already, but we cannot be sure yet."

Concerns Over Stated Reserves

[World] reserves are confused and in fact inflated. Many of the so-called reserves are in fact resources. They're not delineated, they're not accessible, they're not available for production.

—Sadad I. Al-Husseini, former VP of Aramco, presentation to the Oil and Money conference, October 2007.

Al-Husseini estimated that 300 billion barrels ($48{\times}10^9$ m³) of the world's 1,200 billion barrels ($190{\times}10^9$ m³) of proven reserves should be recategorized as speculative resources.

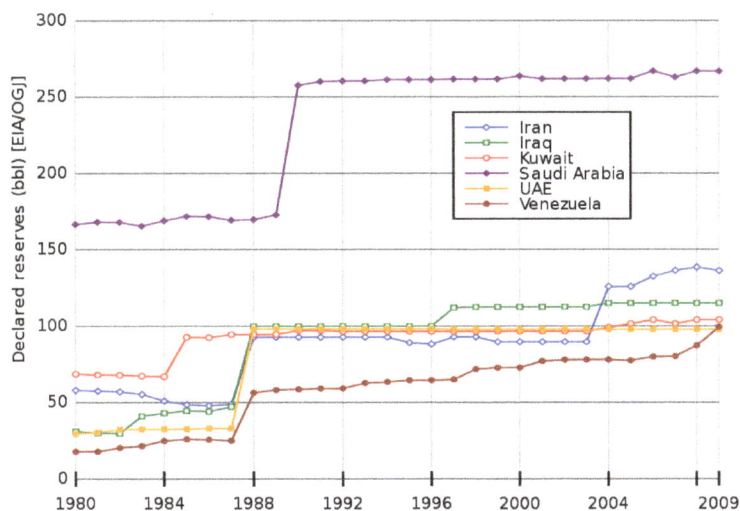

Graph of OPEC reported reserves showing jumps in stated reserves without associated discoveries, as well as the lack of depletion despite yearly production

One difficulty in forecasting the date of peak oil is the opacity surrounding the oil reserves classified as "proven". In many major producing countries, the majority of reserves claims have not been subject to outside audit or examination. Many worrying signs concerning the depletion of proven reserves have emerged in recent years. This was best exemplified by the 2004 scandal surrounding the "evaporation" of 20% of Shell's reserves.

For the most part, proven reserves are stated by the oil companies, the producer states and the consumer states. All three have reasons to overstate their proven reserves: oil companies may look to increase their potential worth; producer countries gain a stronger international stature; and governments of consumer countries may seek a means to foster sentiments of security and stability within their economies and among consumers.

Major discrepancies arise from accuracy issues with the self-reported numbers from the Organization of the Petroleum Exporting Countries (OPEC). Besides the possibility that these nations have overstated their reserves for political reasons (during periods of no substantial discoveries), over 70 nations also follow a practice of not reducing their reserves to account for yearly production. Analysts have suggested that OPEC member nations have economic incentives to exaggerate their reserves, as the OPEC quota system allows greater output for countries with greater reserves.

Kuwait, for example, was reported in the January 2006 issue of *Petroleum Intelligence Weekly* to have only 48 billion barrels ($7.6{\times}10^9$ m³) in reserve, of which only 24 were fully proven. This report was based on the leak of a confidential document from Kuwait and has not been formally denied by the Kuwaiti authorities. This leaked document is from 2001, but excludes revisions or discoveries made since then. Additionally, the reported 1.5 billion barrels ($240{\times}10^6$ m³) of oil burned off by Iraqi soldiers in the First Persian Gulf War are conspicuously missing from Kuwait's figures.

On the other hand, investigative journalist Greg Palast argues that oil companies have an interest in making oil look more rare than it is, to justify higher prices. This view is contested by ecological journalist Richard Heinberg. Other analysts argue that oil producing countries understate the extent of their reserves to drive up the price.

The EUR reported by the 2000 USGS survey of 2,300 billion barrels (370×10⁹ m³) has been criticized for assuming a discovery trend over the next twenty years that would reverse the observed trend of the past 40 years. Their 95% confidence EUR of 2,300 billion barrels (370×10⁹ m³) assumed that discovery levels would stay steady, despite the fact that new-field discovery rates have declined since the 1960s. That trend of falling discoveries has continued in the ten years since the USGS made their assumption. The 2000 USGS is also criticized for other assumptions, as well as assuming 2030 production rates inconsistent with projected reserves.

Reserves of Unconventional Oil

Syncrude's Mildred Lake mine site and plant near Fort McMurray, Alberta

As conventional oil becomes less available, it can be replaced with production of liquids from unconventional sources such as tight oil, oil sands, ultra-heavy oils, gas-to-liquid technologies, coal-to-liquid technologies, biofuel technologies, and shale oil. In the 2007 and subsequent International Energy Outlook editions, the word "Oil" was replaced with "Liquids" in the chart of world energy consumption. In 2009 biofuels was included in "Liquids" instead of in "Renewables". The inclusion of natural gas liquids, a bi-product of natural gas extraction, in "Liquids" has been criticized as it is mostly a chemical feedstock which is generally not used as transport fuel.

Texas oil production declined since peaking in 1972 but has recently had a resurgence due to tight oil production

Reserve estimates are based on the oil price. Hence, unconventional sources such as heavy crude oil, oil sands, and oil shale may be included as new techniques reduce the cost of extraction. With rule changes by the SEC, oil companies can now book them as proven reserves after opening a strip mine or thermal facility for extraction. These unconventional sources are more labor and resource intensive to produce, however, requiring extra energy to refine, resulting in higher production costs and up to three times more greenhouse gas emissions per barrel (or barrel equivalent) on a "well to tank" basis or 10 to 45% more on a "well to wheels" basis, which includes the carbon emitted from combustion of the final product.

While the energy used, resources needed, and environmental effects of extracting unconventional sources have traditionally been prohibitively high, major unconventional oil sources being considered for large-scale production are the extra heavy oil in the Orinoco Belt of Venezuela, the Athabasca Oil Sands in the Western Canadian Sedimentary Basin, and the oil shale of the Green River Formation in Colorado, Utah, and Wyoming in the United States. Energy companies such as Syncrude and Suncor have been extracting bitumen for decades but production has increased greatly in recent years with the development of Steam Assisted Gravity Drainage and other extraction technologies.

Chuck Masters of the USGS estimates that, "Taken together, these resource occurrences, in the Western Hemisphere, are approximately equal to the Identified Reserves of conventional crude oil accredited to the Middle East." Authorities familiar with the resources believe that the world's ultimate reserves of unconventional oil are several times as large as those of conventional oil and will be highly profitable for companies as a result of higher prices in the 21st century. In October 2009, the USGS updated the Orinoco tar sands (Venezuela) recoverable "mean value" to 513 billion barrels (8.16×10^{10} m³), with a 90% chance of being within the range of 380-652 billion barrels (103.7×10^9 m³), making this area "one of the world's largest recoverable oil accumulations".

Total World Oil Reserves

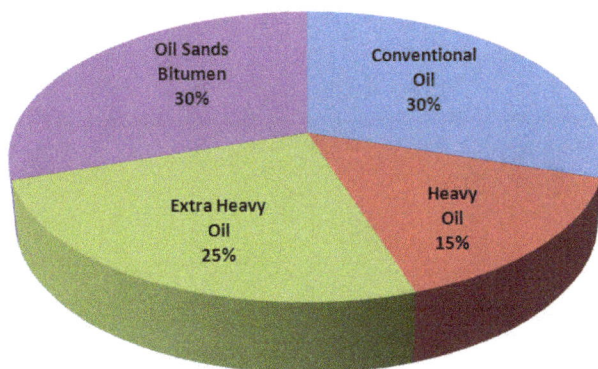

Unconventional resources are much larger than conventional ones

Despite the large quantities of oil available in non-conventional sources, Matthew Simmons argued in 2005 that limitations on production prevent them from becoming an effective substitute for conventional crude oil. Simmons stated "these are high energy intensity projects that can never reach high volumes" to offset significant losses from other sources. Another study claims that even under highly optimistic assumptions, "Canada's oil sands will not prevent peak oil," although production could reach 5,000,000 bbl/d (790,000 m³/d) by 2030 in a "crash program" development effort.

Moreover, oil extracted from these sources typically contains contaminants such as sulfur and heavy metals that are energy-intensive to extract and can leave tailings, ponds containing hydrocarbon sludge, in some cases. The same applies to much of the Middle East's undeveloped conventional oil reserves, much of which is heavy, viscous, and contaminated with sulfur and metals to the point of being unusable. However, high oil prices make these sources more financially appealing. A study by Wood Mackenzie suggests that by the early 2020s all the world's extra oil supply is likely to come from unconventional sources.

Production

The point in time when peak global oil production occurs defines peak oil. Some adherents of 'peak oil' believe that production capacity will remain the main limitation of supply, and that when production decreases, it will be the main bottleneck to the petroleum supply/demand equation. Others believe that the increasing industrial effort to extract oil will have a negative effect on global economic growth, leading to demand contraction and a price collapse, thereby causing production decline as some unconventional sources become uneconomical. Yet others believe that the peak may be to some extent led by declining demand as new technologies and improving efficiency shift energy usage away from oil.

Worldwide oil discoveries have been less than annual production since 1980. World population has grown faster than oil production. Because of this, oil production *per capita* peaked in 1979 (preceded by a plateau during the period of 1973–1979).

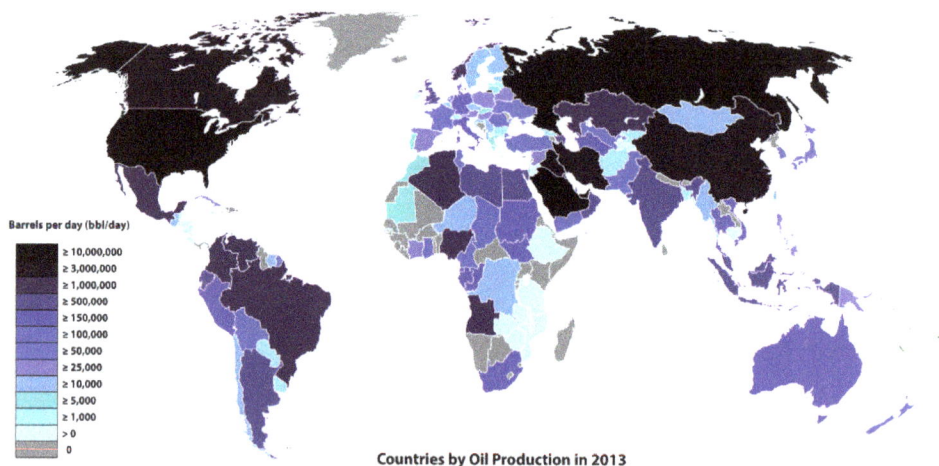

Countries producing oil 2013, bbl/day (CIA World Factbook)

The increasing investment in harder-to-reach oil is a sign of oil companies' belief in the end of easy oil. Also, while it is widely believed that increased oil prices spur an increase in production, an increasing number of oil industry insiders were reportedly coming to believe that even with higher prices, oil production was unlikely to increase significantly. Among the reasons cited were both geological factors as well as "above ground" factors that are likely to see oil production plateau.

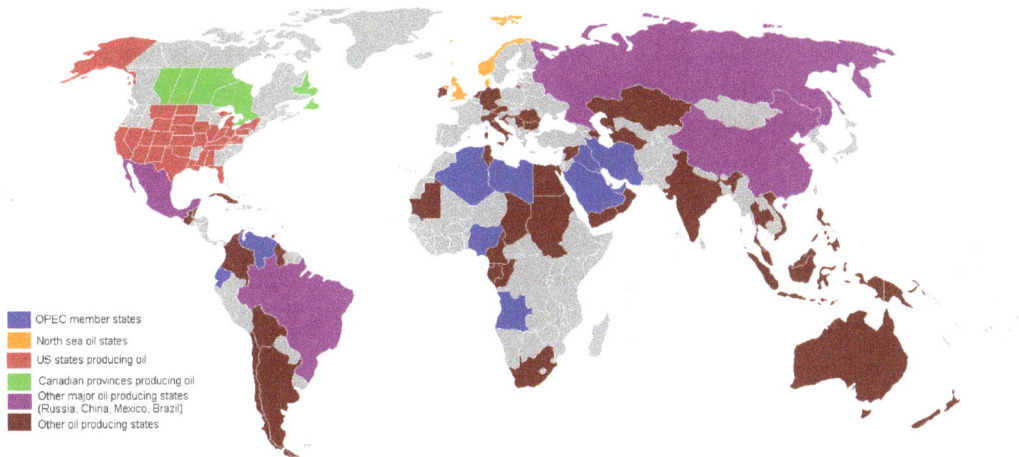

Oil producing countries

An important concept with regard to declining "easy oil" is energy returned on energy invested, also referred to as EROEI. A 2008 *Journal of Energy Security* analysis of the energy return on drilling effort in the United States concluded that there was extremely limited potential to increase production of both gas and (especially) oil. By looking at the historical response of production to variation in drilling effort, the analysis showed very little increase of production attributable to increased drilling. This was because of a tight quantitative relationship of diminishing returns with increasing drilling effort: as drilling effort increased, the energy obtained per active drill rig was reduced according to a severely diminishing power law. The study concluded that even an enormous increase of drilling effort was unlikely to significantly increase oil and gas production in a mature petroleum region such as the United States. However, contrary to the study's conclusion, since the analysis was published in 2008, US production of crude oil has increased 74%, and production of dry natural gas has increased 28% (2014 compared to 2008).

Anticipated Production by Major Agencies

Who exported Petroleum oils, crude in 2012?

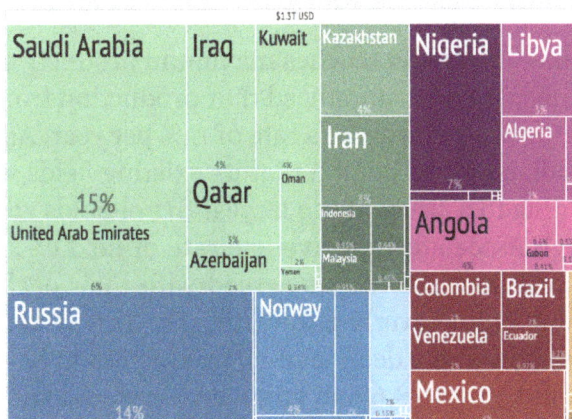

Crude oil export treemap (2012) from Harvard Atlas of Economic Complexity

Average yearly gains in global supply from 1987 to 2005 were 1.2 million barrels per day (190×10³ m³/d) (1.7%). In 2005, the IEA predicted that 2030 production rates would reach

120,000,000 barrels per day (19,000,000 m³/d), but this number was gradually reduced to 105,000,000 barrels per day (16,700,000 m³/d). A 2008 analysis of IEA predictions questioned several underlying assumptions and claimed that a 2030 production level of 75,000,000 ba rels per day (11,900,000 m³/d) (comprising 55,000,000 barrels (8,700,000 m³) of crude oil and 20,000,000 barrels (3,200,000 m³) of both non-conventional oil and natural gas liquids) was more realistic than the IEA numbers. More recently, the EIA's Annual Energy Outlook 2015 indicated no production peak out to 2040. However, this required a future Brent crude oil price of $US144/bbl (2013 dollars) "as growing demand leads to the development of more costly resources." Whether the world economy can grow and maintain demand for such a high oil price remains to be seen.

Oil Field Decline

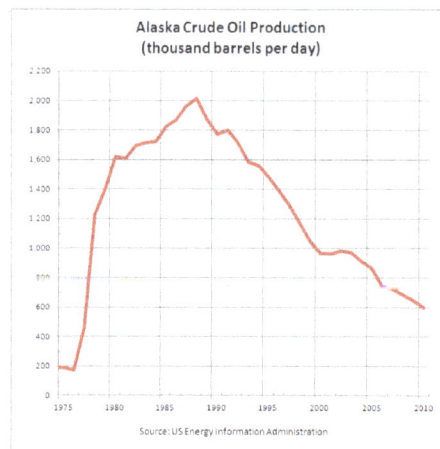

Alaska's oil production has declined 70% since peaking in 1988

In a 2013 study of 733 giant oil fields, only 32% of the ultimately recoverable oil, condensate and gas remained. Ghawar, which is the largest oil field in the world and responsible for approximately half of Saudi Arabia's oil production over the last 50 years, was in decline before 2009. The world's second largest oil field, the Burgan Field in Kuwait, entered decline in November 2005.

It is well established that once an oilfield reaches maximum production, it will decrease at a certain decline rate. For example, Mexico announced that production from its giant Cantarell Field began to decline in March 2006, reportedly at a rate of 13% per year. Also in 2006, Saudi Aramco Senior Vice President Abdullah Saif estimated that its existing fields were declining at a rate of 5% to 12% per year. According to a study of the largest 811 oilfields conducted in early 2008 by Cambridge Energy Research Associates, the average rate of field decline is 4.5% per year. The Association for the Study of Peak Oil and Gas agreed with their decline rates, but considered the rate of new fields coming online overly optimistic. The IEA stated in November 2008 that an analysis of 800 oilfields showed the decline in oil production to be 6.7% a year for fields past their peak, and that this would grow to 8.6% in 2030. A more rapid annual rate of decline of 5.1% in 800 of the world's largest oil fields weighted for production over their whole lives was reported by the International Energy Agency in their World Energy Outlook 2008. The 2013 study of 733 giant fields mentioned previously had an average decline rate 3.83% which was described as "conservative."

Control Over Supply

Entities such as governments or cartels can reduce supply to the world market by limiting access to the supply through nationalizing oil, cutting back on production, limiting drilling rights, imposing taxes, etc. International sanctions, corruption, and military conflicts can also reduce supply.

Nationalization of Oil Supplies

Another factor affecting global oil supply is the nationalization of oil reserves by producing nations. The nationalization of oil occurs as countries begin to deprivatize oil production and withhold exports. Kate Dourian, Platts' Middle East editor, points out that while estimates of oil reserves may vary, politics have now entered the equation of oil supply. "Some countries are becoming off limits. Major oil companies operating in Venezuela find themselves in a difficult position because of the growing nationalization of that resource. These countries are now reluctant to share their reserves."

According to consulting firm PFC Energy, only 7% of the world's estimated oil and gas reserves are in countries that allow companies like ExxonMobil free rein. Fully 65% are in the hands of state-owned companies such as Saudi Aramco, with the rest in countries such as Russia and Venezuela, where access by Western European and North American companies is difficult. The PFC study implies political factors are limiting capacity increases in Mexico, Venezuela, Iran, Iraq, Kuwait, and Russia. Saudi Arabia is also limiting capacity expansion, but because of a self-imposed cap, unlike the other countries. As a result of not having access to countries amenable to oil exploration, ExxonMobil is not making nearly the investment in finding new oil that it did in 1981.

OPEC Influence on Supply

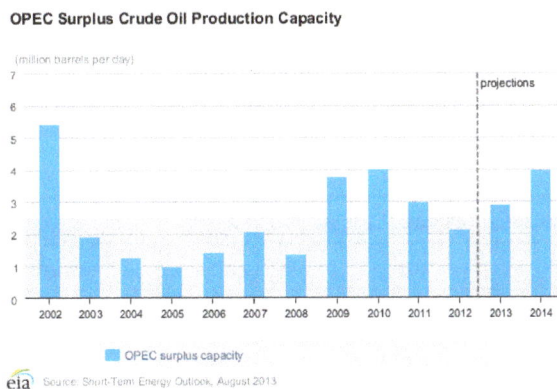

OPEC surplus crude oil production capacity (US EIA)

OPEC is an alliance between 12 diverse oil producing countries (Algeria, Angola, Ecuador, Iran, Iraq, Kuwait, Libya, Nigeria, Qatar, Saudi Arabia, the United Arab Emirates, and Venezuela) to control the supply of oil. OPEC's power was consolidated as various countries nationalized their oil holdings, and wrested decision-making away from the "Seven Sisters," (Anglo-Iranian, Socony-Vacuum, Royal Dutch Shell, Gulf, Esso, Texaco, and Socal) and created their own oil companies to control the oil. OPEC tries to influence prices by restricting production. It does this by allocating each member country a quota for production. All 12 members agree to keep prices high by producing at lower levels than they otherwise would. There is no way to verify adherence to the quota, so every member faces the same incentive to "cheat" the cartel. United States policy of selling arms and providing security to Saudi Arabia is often seen as an attempt to influence the Saudis to increase oil production. According to sociology professor Michael Schwartz, the purpose for the second Iraq war was to break the back of OPEC and return control of the oil fields to western oil companies.

Alternatively, commodities trader Raymond Learsy, author of *Over a Barrel: Breaking the Middle East Oil Cartel*, contends that OPEC has trained consumers to believe that oil is a much more finite resource than it is. To back his argument, he points to past false alarms and apparent collaboration. He also believes that peak oil analysts are conspiring with OPEC and the oil companies to create a "fabricated drama of peak oil" to drive up oil prices and profits; oil had risen to a little over $30/barrel at that time. A counter-argument was given in the Huffington Post after he and Steve Andrews, co-founder of ASPO, debated on CNBC in June 2007.

Predictions of Peak Oil

Pub.	Made by	Peak year/range	Pub.	Made by	Peak year/range
1972	Esso	About 2000	1999	Parker	2040
1972	United Nations	By 2000	2000	A. A. Bartlett	2004 or 2019
1974	Hubbert	1991–2000	2000	Duncan	2006
1976	UK Dep. of Energy	About 2000	2000	EIA	2021–2067; 2037 most likely
1977	Hubbert	1996	2000	EIA (WEO)	Beyond 2020
1977	Ehrlich, et al.	2000	2001	Deffeyes	2003–2008
1979	Shell	Plateau by 2004	2001	Goodstein	2007
1981	World Bank	Plateau around 2000	2002	Smith	2010–2016
1985	J. Bookout	2020	2002	Campbell	2010
1989	Campbell	1989	2002	Cavallo	2025–2028
1994	L. F. Ivanhoe	OPEC plateau 2000–2050	2003	Greene, et al.	2020–2050
1995	Petroconsultants	2005	2003	Laherrère	2010–2020
1997	Ivanhoe	2010	2003	Lynch	No visible peak
1997	J. D. Edwards	2020	2003	Shell	After 2025
1998	IEA	2014	2003	Simmons	2007–2009
1998	Campbell & Laherrère	2004	2004	Bakhitari	2006–2007
1999	Campbell	2010	2004	CERA	After 2020
1999	Peter Odell	2060	2004	PFC Energy	2015–2020

A selection of estimates of the year of peak world oil production, compiled by the United States Energy Information Administration

In 1962, Hubbert predicted that world oil production would peak at a rate of 12.5 billion barrels per year, around the year 2000. In 1974, Hubbert predicted that peak oil would occur in 1995 "if current trends continue." Those predictions proved incorrect. However, a number of industry leaders and analysts believe that world oil production will peak between 2015 and 2030, with a significant chance that the peak will occur before 2020. They consider dates after 2030 implausible. By comparison, a 2014 analysis of production and reserve data predicted a peak in oil production about 2035. Determining a more specific range is difficult due to the lack of certainty over the actual size of world oil reserves. Unconventional oil is not currently predicted to meet the expected shortfall even in a best-case scenario. For unconventional oil to fill the gap without "potentially serious impacts on the global economy", oil production would have to remain stable after its peak, until 2035 at the earliest.

Papers published since 2010 have been relatively pessimistic. A 2010 Kuwait University study predicted production would peak in 2014. A 2010 Oxford University study predicted that production will peak before 2015, but its projection of a change soon "... from a demand-led market to a supply constrained market ..." was incorrect. A 2014 validation of a significant 2004 study in the journal *Energy* proposed that it is likely that conventional oil production peaked, according to various definitions, between 2005 and 2011. A set of models published in a 2014 Ph.D. thesis predicted that a 2012 peak would be followed by a drop in oil prices, which in some scenarios could turn into a rapid rise in prices thereafter. According to energy blogger Ron Patterson, the peak of world oil production was probably around 2010.

Major oil companies hit peak production in 2005. Several sources in 2006 and 2007 predicted that worldwide production was at or past its maximum. Fatih Birol, chief economist at the International Energy Agency, also stated that "crude oil production for the world has already peaked in 2006." However, in 2013 OPEC's figures showed that world crude oil production and remaining proven reserves were at record highs. According to Matthew Simmons, former Chairman of Simmons & Company International and author of *Twilight in the Desert: The Coming Saudi Oil Shock and the World Economy*, "peaking is one of these fuzzy events that you only know clearly when you see it through a rear view mirror, and by then an alternate resolution is generally too late."

Possible Consequences

The wide use of fossil fuels has been one of the most important stimuli of economic growth and prosperity since the industrial revolution, allowing humans to participate in takedown, or the consumption of energy at a greater rate than it is being replaced. Some believe that when oil production decreases, human culture, and modern technological society will be forced to change drastically. The impact of peak oil will depend heavily on the rate of decline and the development and adoption of effective alternatives.

In 2005, the United States Department of Energy published a report titled *Peaking of World Oil Production: Impacts, Mitigation, & Risk Management*. Known as the Hirsch report, it stated, "The peaking of world oil production presents the U.S. and the world with an unprecedented risk management problem. As peaking is approached, liquid fuel prices and price volatility will increase dramatically, and, without timely mitigation, the economic, social, and political costs will be unprecedented. Viable mitigation options exist on both the supply and demand sides, but to have substantial impact, they must be initiated more than a decade in advance of peaking." Some of the information was updated in 2007.

Oil Prices

Historical Oil Prices

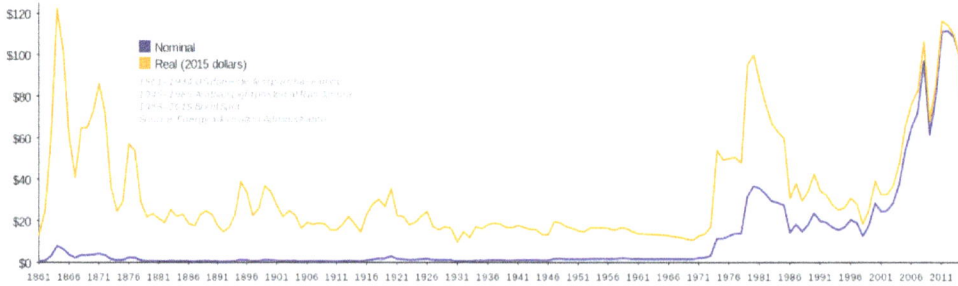

Long-term oil prices, 1861–2015 (top line adjusted for inflation)

The oil price historically was comparatively low until the 1973 oil crisis and the 1979 energy crisis when it increased more than tenfold during that six-year timeframe. Even though the oil price dropped significantly in the following years, it has never come back to the previous levels. Oil price began to increase again during the 2000s until it hit historical heights of $143 per barrel (2007 inflation adjusted dollars) on 30 June 2008. As these prices were well above those that caused the 1973 and 1979 energy crises, they contributed to fears of an economic recession similar to that of the early 1980s.

It is generally agreed that the main reason for the price spike in 2005–2008 was strong demand pressure. For example, global consumption of oil rose from 30 billion barrels (4.8×10^9 m³) in 2004 to 31 billion in 2005. The consumption rates were far above new discoveries in the period, which had fallen to only eight billion barrels of new oil reserves in new accumulations in 2004.

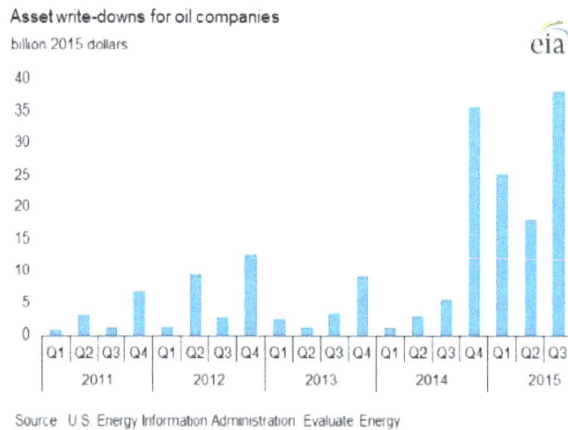

Asset write downs for oil companies 2015

Oil price increases were partially fueled by reports that petroleum production is at or near full capacity. In June 2005, OPEC stated that they would 'struggle' to pump enough oil to meet pricing pressures for the fourth quarter of that year. From 2007 to 2008, the decline in the U.S. dollar against other significant currencies was also considered as a significant reason for the oil price increases, as the dollar lost approximately 14% of its value against the Euro from May 2007 to May 2008.

Besides supply and demand pressures, at times security related factors may have contributed to increases in prices, including the War on Terror, missile launches in North Korea, the Crisis between Israel and Lebanon, nuclear brinkmanship between the U.S. and Iran, and reports from the U.S. Department of Energy and others showing a decline in petroleum reserves.

West Texas Intermediate (WTI) Crude Oil Price

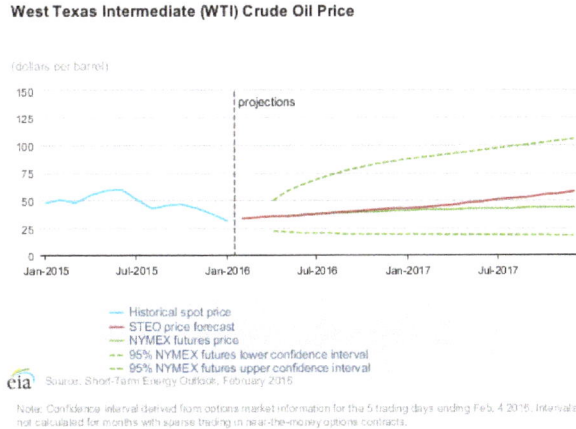

Depicts EIA projections for West Texas Intermediate crude oil price for 2016-2017

More recently, between 2011 and 2014 the price of crude oil was relatively stable, fluctuating around $US100 per barrel. It dropped sharply in late 2014 to below $US70 where it remained for most of 2015. In early 2016 it traded at a low of $US27. The price drop has been attributed to both oversupply and reduced demand as a result of the slowing global economy, OPEC reluctance to concede market share, and a stronger US dollar. These factors may be exacerbated by a combination of monetary policy and the increased debt of oil producers, who may increase production to maintain liquidity.

This price drop has placed many US tight oil producers under considerable financial pressure. As a result, there has been a reduction by oil companies in capital expenditure of over $US400 billion. It is anticipated that this will have effects on global production in the longer term, leading to statements of concern by the International Energy Agency that governments should not be complacent about energy security. Energy Information Agency projections anticipate market oversupply and prices below $US50 until late 2017.

Effects of Historical Oil Price Rises

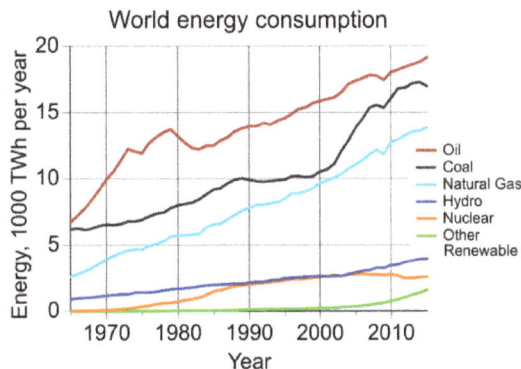

World consumption of primary energy by energy type

In the past, the price of oil has led to economic recessions, such as the 1973 and 1979 energy crises. The effect the price of oil has on an economy is known as a price shock. In many European countries, which have high taxes on fuels, such price shocks could potentially be mitigated somewhat by temporarily or permanently suspending the taxes as fuel costs rise. This method of softening price shocks is less useful in countries with much lower gas taxes, such as the United States. A baseline scenario for a recent IMF paper found oil production growing at 0.8% (as opposed to a historical average of 1.8%) would result in a small reduction in economic growth of 0.2–0.4%.

Researchers at the Stanford Energy Modeling Forum found that the economy can adjust to steady, gradual increases in the price of crude better than wild lurches.

Some economists predict that a substitution effect will spur demand for alternate energy sources, such as coal or liquefied natural gas. This substitution can be only temporary, as coal and natural gas are finite resources as well.

Prior to the run-up in fuel prices, many motorists opted for larger, less fuel-efficient sport utility vehicles and full-sized pickups in the United States, Canada, and other countries. This trend has been reversing because of sustained high prices of fuel. The September 2005 sales data for all vehicle vendors indicated SUV sales dropped while small cars sales increased. Hybrid and diesel vehicles are also gaining in popularity.

EIA published Household Vehicles Energy Use: Latest Data and Trends in Nov 2005 illustrating the steady increase in disposable income and $20–30 per barrel price of oil in 2004. The report notes "The average household spent $1,520 on fuel purchases for transport." According to CNBC that expense climbed to $4,155 in 2011.

In 2008, a report by Cambridge Energy Research Associates stated that 2007 had been the year of peak gasoline usage in the United States, and that record energy prices would cause an "enduring shift" in energy consumption practices. The total miles driven in the U.S. peaked in 2006.

The Export Land Model states that after peak oil petroleum exporting countries will be forced to reduce their exports more quickly than their production decreases because of internal demand growth. Countries that rely on imported petroleum will therefore be affected earlier and more dramatically than exporting countries. Mexico is already in this situation. Internal consumption grew by 5.9% in 2006 in the five biggest exporting countries, and their exports declined by over 3%. It was estimated that by 2010 internal demand would decrease worldwide exports by 2,500,000 barrels per day (400,000 m³/d).

Canadian economist Jeff Rubin has stated that high oil prices are likely to result in increased consumption in developed countries through partial manufacturing de-globalisation of trade. Manufacturing production would move closer to the end consumer to minimise transportation network costs, and therefore a demand decoupling from gross domestic product would occur. Higher oil prices would lead to increased freighting costs and consequently, the manufacturing industry would move back to the developed countries since freight costs would outweigh the current economic wage advantage of developing countries. Economic research carried out by the International Monetary Fund puts overall price elasticity of demand for oil at −0.025 short-term and −0.093 long term.

Agricultural Effects and Population Limits

Since supplies of oil and gas are essential to modern agriculture techniques, a fall in global oil supplies could cause spiking food prices and unprecedented famine in the coming decades.[note 1] Geologist Dale Allen Pfeiffer contends that current population levels are unsustainable, and that to achieve a sustainable economy and avert disaster the United States population would have to be reduced by at least one-third, and world population by two-thirds.

The largest consumer of fossil fuels in modern agriculture is ammonia production (for fertilizer) via the Haber process, which is essential to high-yielding intensive agriculture. The specific fossil fuel input to fertilizer production is primarily natural gas, to provide hydrogen via steam reforming. Given sufficient supplies of renewable electricity, hydrogen can be generated without fossil fuels using methods such as electrolysis. For example, the Vemork hydroelectric plant in Norway used its surplus electricity output to generate renewable ammonia from 1911 to 1971.

Iceland currently generates ammonia using the electrical output from its hydroelectric and geothermal power plants, because Iceland has those resources in abundance while having no domestic hydrocarbon resources, and a high cost for importing natural gas.

Long-term Effects on Lifestyle

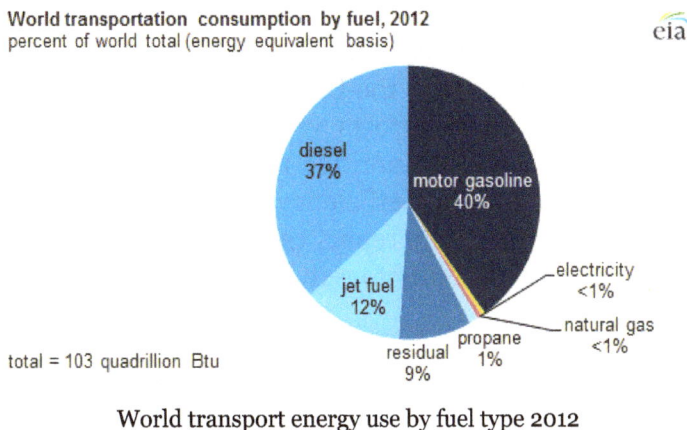

World transport energy use by fuel type 2012

A majority of Americans live in suburbs, a type of low-density settlement designed around universal personal automobile use. Commentators such as James Howard Kunstler argue that because over 90% of transportation in the U.S. relies on oil, the suburbs' reliance on the automobile is an unsustainable living arrangement. Peak oil would leave many Americans unable to afford petroleum based fuel for their cars, and force them to use bicycles or electric vehicles. Additional options include telecommuting, moving to rural areas, or moving to higher density areas, where walking and public transportation are more viable options. In the latter two cases, suburbs may become the "slums of the future." The issue of petroleum supply and demand is also a concern for growing cities in developing countries (where urban areas are expected to absorb most of the world's projected 2.3 billion population increase by 2050). Stressing the energy component of future development plans is seen as an important goal.

Rising oil prices, if they occur, would also affect the cost of food, heating, and electricity. A high amount of stress would then be put on current middle to low income families as economies contract from the decline in excess funds, decreasing employment rates. The Hirsch/US DoE Report concludes that "without timely mitigation, world supply/demand balance will be achieved through massive demand destruction (shortages), accompanied by huge oil price increases, both of which would create a long period of significant economic hardship worldwide."

Methods that have been suggested for mitigating these urban and suburban issues include the use of non-petroleum vehicles such as electric cars, battery electric vehicles, transit-oriented development, carfree cities, bicycles, new trains, new pedestrianism, smart growth, shared space, urban consolidation, urban villages, and New Urbanism.

An extensive 2009 report on the effects of compact development by the United States National Research Council of the Academy of Sciences, commissioned by the United States Congress, stated six main findings. First, that compact development is likely to reduce "Vehicle Miles Traveled" (VMT) throughout the country. Second, that doubling residential density in a given area could reduce VMT by as much as 25% if coupled with measures such as increased employment density and improved public transportation. Third, that higher density, mixed-use developments would produce both direct reductions in CO_2 emissions (from less driving), and indirect reductions (such as from lower amounts of materials used per housing unit, higher efficiency climate control, longer vehicle lifespans, and higher efficiency delivery of goods and services). Fourth, that although short term reductions in energy use and CO_2 emissions would be modest, that these reductions would become more significant over time. Fifth, that a major obstacle to more compact development in the United States is political resistance from local zoning regulators, which would hamper efforts by state and regional governments to participate in land-use planning. Sixth, the committee agreed that changes in development that would alter driving patterns and building efficiency would have various secondary costs and benefits that are difficult to quantify. The report recommends that policies supporting compact development (and especially its ability to reduce driving, energy use, and CO_2 emissions) should be encouraged.

An economic theory that has been proposed as a remedy is the introduction of a steady state economy. Such a system could include a tax shifting from income to depleting natural resources (and pollution), as well as the limitation of advertising that stimulates demand and population growth. It could also include the institution of policies that move away from globalization and toward localization to conserve energy resources, provide local jobs, and maintain local decision-making authority. Zoning policies could be adjusted to promote resource conservation and eliminate sprawl.

Mitigation

To avoid the serious social and economic implications a global decline in oil production could entail, the Hirsch report emphasized the need to find alternatives, at least ten to twenty years before the peak, and to phase out the use of petroleum over that time. This was similar to a plan proposed for Sweden that same year. Such mitigation could include energy conservation, fuel substitution, and the use of unconventional oil. The timing of mitigation responses is critical. Premature initiation would be undesirable, but if initiated too late could be more costly and have more negative economic consequences.

Positive Aspects

Permaculture sees peak oil as holding tremendous potential for positive change, assuming countries act with foresight. The rebuilding of local food networks, energy production, and the general implementation of "energy descent culture" are argued to be ethical responses to the acknowledgment of finite fossil resources. Majorca is an island currently diversifying its energy supply from fossil fuels to alternative sources and looking back at traditional construction and permaculture methods.

The Transition Towns movement, started in Totnes, Devon and spread internationally by "The Transition Handbook" (Rob Hopkins) and Transition Network, sees the restructuring of society for more local resilience and ecological stewardship as a natural response to the combination of peak oil and climate change.

Criticisms

General Arguments

Opponents to the theory of peak oil often cite new oil reserves that have been found, which continue to forestall a peak oil event. In particular, some contend that oil production from these new oil reserves as well as from existing fields will continue to increase at a rate that outpaces demand, until alternate energy sources for our current fossil fuel dependence are found. As of 2015, analysts in both the petroleum and financial industries were concluding that the "age of oil" had already reached a new stage where the excess supply that appeared in late 2014 may continue to prevail in the future. A consensus appeared to be emerging that an international agreement will be reached to introduce measures to constrain the combustion of hydrocarbons in an effort to limit global temperature rise to the nominal 2 °C that is consensually predicted to limit environmental harm to tolerable levels.

Further criticism against peak oil is confidence in the various options and technologies for substituting oil. And indeed there are some promising approaches that seem to have the potential to reduce or even counterbalance the effects of a peak oil situation. For example, US federal funding has increased for algae fuels since the year 2000 due to rising fuel prices. Numerous more projects are being funded in Australia, New Zealand, Europe, the Middle East, and other parts of the world and private companies are entering the field.

Oil Industry Representatives

Oil industry representatives have criticised peak oil theory, at least as it has been presented by Matthew Simmons. The president of Royal Dutch Shell's U.S. operations John Hofmeister, while agreeing that conventional oil production would soon start to decline, criticized Simmons's analysis for being "overly focused on a single country: Saudi Arabia, the world's largest exporter and OPEC swing producer." He also pointed to the large reserves at the US outer continental shelf, which held an estimated 100 billion barrels (16×10^9 m³) of oil and natural gas. However, only 15% of those reserves were currently exploitable, a good part of that off the coasts of Louisiana, Alabama, Mississippi, and Texas. Hofmeister also contended that Simmons erred in excluding unconventional sources of oil such as the oil sands of Canada, where Shell was active. The Canadian oil sands—a natural combination of sand, water, and oil found largely in Alberta and Saskatchewan—are believed to contain one trillion barrels of oil. Another trillion barrels are also said to be trapped in rocks in Colorado, Utah, and Wyoming, but are in the form of oil shale. These particular reserves present major environmental, social, and economic obstacles to recovery. Hofmeister also claimed that if oil companies were allowed to drill more in the United States enough to produce another 2 million barrels per day (320×10^3 m³/d), oil and gas prices would not be as high as they were in the later part of the 2000 to 2010 decade. He thought in 2008 that high energy prices would cause social unrest similar to the 1992 Rodney King riots.

In 2009, Dr. Christoph Rühl, chief economist of BP, argued against the peak oil hypothesis:

Physical peak oil, which I have no reason to accept as a valid statement either on theoretical, scientific or ideological grounds, would be insensitive to prices. ... In fact the whole hypothesis of peak oil – which is that there is a certain amount of oil in the ground, consumed at a certain rate, and then it's finished – does not react to anything ... Therefore there will never be a moment when the world runs out of oil because there will always be a price at which the last drop of oil can clear the market. And you can turn anything into oil into if you are willing to pay the financial and environmental price ... (Global Warming) is likely to be more of a natural limit than all these peak oil theories combined. ... Peak oil has been predicted for 150 years. It has never happened, and it will stay this way.

According to Rühl, the main limitations for oil availability are "above ground" and are to be found in the availability of staff, expertise, technology, investment security, money and last but not least in global warming. The oil question is about price and not the basic availability. Rühl's views are shared by Daniel Yergin of CERA, who added that the recent high price phase might add to a future demise of the oil industry, not of complete exhaustion of resources or an apocalyptic shock but the timely and smooth setup of alternatives.

Clive Mather, CEO of Shell Canada, said the Earth's supply of bitumen hydrocarbons is "almost infinite", referring to hydrocarbons in oil sands.

Others

Economist Robert L. Bradley, Jr. wrote in a 2007 article in *The Review of Austrian Economics* that, "[a]n Austrian institutional theory is more robust for explaining changes in mineral-resource scarcity than neoclassical depletionism[.]" Using the writings of Erich Zimmermann and Julian Simon, Bradley also argued in 2012 that resources have subjective rather than objective existences in economics. He concluded that, "what resources come from the ground ultimately depend on the resources in the mind."

Attorney and mechanical engineer Peter W. Huber pointed out in 2006 that the world is just running out of "cheap oil." As oil prices rise, unconventional sources become economically viable. He predicted that, "[t]he tar sands of Alberta alone contain enough hydrocarbon to fuel the entire planet for over 100 years."

Industry blogger Steve Maley echoed some of the points of Yergin, Rühl, Mather and Hofmeister.

Environmental journalist George Monbiot responded to a 2012 report by Leonardo Maugeri by proclaiming that there is more than enough oil (from unconventional sources) for capitalism to "deep-fry" the world with climate change. Stephen Sorrell, senior lecturer Science and Technology Policy Research, Sussex Energy Group, and lead author of the UKERC Global Oil Depletion report, and Christophe McGlade, doctoral researcher at the UCL Energy Institute have criticized Maugeri's assumptions about decline rates.

Hirsch Report

The Hirsch report, the commonly referred to name for the report Peaking of World Oil Production: Impacts, Mitigation, and Risk Management, was created by request for the US Department of Energy and published in February 2005. Some information was updated in 2007. It examined the time frame for the occurrence of peak oil, the necessary mitigating actions, and the likely impacts based on the timeliness of those actions.

Introduction

"The peaking of world oil production presents the U.S. and the world with an unprecedented risk management problem. As peaking is approached, liquid fuel prices and price volatility will increase dramatically, and, without timely mitigation, the economic, social, and political costs will be unprecedented. Viable mitigation options exist on both the supply and demand sides, but to have substantial impact, they must be initiated more than a decade in advance of peaking."

The lead author, Robert L. Hirsch, published a brief summary of this report in October 2005 for the Atlantic Council.

Projections

A number of industry petroleum geologists, scientists, and economists were listed with their global peak production projection. Later, in 2010, Hirsch developed a projection that global oil production would begin to decline by 2015.

Projected Date	Source
2006–2007	Bakhtiari
2007–2009	Simmons
After 2007	Skrebowski
Before 2009	Deffeyes
Before 2010	Goodstein
Around 2010	Campbell
After 2010	World Energy Council
2010–2020	Laherrere
2016	EIA (Nominal)
After 2020	CERA
2025 or later	Shell

Mitigation

Operating under the assumption that existing services must be sustained, the Hirsch report considered the effects of the following mitigation strategies as part of the "crash program":

1. Fuel efficient transportation,

2. Heavy oil/Oil sands,

3. Coal liquefaction,

4. Enhanced oil recovery,

5. Gas-to-liquids.

Conclusions

The report came to the following conclusions:

- World oil peaking is going to happen, and will likely be abrupt.

    o World production of conventional oil will reach a maximum and decline thereafter.

    o Some forecasters project peaking within a decade; others contend it will occur later.

    o Peaking will happen, but the timing is uncertain.

- Oil peaking will adversely affect global economies, particularly the U.S.

    o Over the past century, the U.S. economy has been shaped by the availability of low-cost oil.

    o The economic loss to the United States could be measured on a trillion-dollar scale.

    o Aggressive fuel efficiency and substitute fuel production could provide substantial mitigation.

- Oil peaking presents a unique challenge.

    o Without massive mitigation, the problem will be pervasive and long-term.

    o Previous energy transitions (wood to coal and coal to oil) were gradual and evolutionary.

    o Oil peaking will be abrupt and revolutionary.

- The problem is liquid fuels for transportation.

    o The lifetimes of transportation equipment are measured in decades.

    o Rapid changeover in transportation equipment is inherently impossible.

    o Motor vehicles, aircraft, trains, and ships have no ready alternative to liquid fuels.

- Mitigation efforts will require substantial time.

    o Waiting until production peaks would leave the world with a liquid fuel deficit for 20 years.

    o Initiating a crash program 10 years before peaking leaves a liquid fuels shortfall of a decade.

    o Initiating a crash program 20 years before peaking could avoid a world liquid fuels shortfall.

- Both supply and demand will require attention.

    o Sustained high oil prices will cause forced demand reduction (recession and unemployment).

    o Production of large amounts of substitute liquid fuels can and must be provided.

    o The production of substitute liquid fuels is technically and economically feasible.

- It is a matter of risk management.

    o The peaking of world oil production is a classic risk management problem.

    o Mitigation efforts earlier than required may be premature, if peaking is long delayed.

    o On the other hand, if peaking is soon, failure to initiate mitigation could be extremely damaging.

- Government intervention will be required.

    o The economic and social implications of oil peaking would otherwise be chaotic.

    o Expediency may require major changes to existing administrative and regulatory procedures.

- Economic upheaval is not inevitable.

    o Without mitigation, the peaking of world oil production will cause major economic upheaval.

    o Given enough lead-time, the problems are soluble with existing technologies.

    o New technologies will help, but on a longer time scale.

- More information is needed.

    o Effective action to combat peaking requires better understanding of the issues.

    o Risks and possible benefits of possible mitigation actions need to be examined.

Three Scenarios

- Waiting until world oil production peaks before taking crash program action leaves the world with a significant liquid fuel deficit for more than two decades.

- Initiating a mitigation crash program 10 years before world oil peaking helps considerably but still leaves a liquid fuels shortfall roughly a decade after the time that oil would have peaked.

- Initiating a mitigation crash program 20 years before peaking appears to offer the possibility of avoiding a world liquid fuels shortfall for the forecast period.

Applicability Beyond the US, Critical Remarks

The Hirsch Report urged a crash program of new technologies and changes in manners and attitudes in the US and as well implying more research and development. The report cites a peaking crude oil supply as the main reason for immediate action.

During the significant oil price rise through 2007, a theme among several industry observers was that the price rise was only partially due to a limit in crude oil availability (peak oil). For example, an article by Jad Mouawad cited an unusual number of fires and other outages among U.S. refineries in the summer of 2007 which disrupted supply. However, a lack of refining capacity would only seem to explain high gasoline prices not high crude oil prices. Indeed, if the refineries were unable to process available crude oil then there should be a crude oil glut that would reduce crude prices on international crude oil markets. Then again, sharp changes in crude oil prices can also be due to stock market volatility and fear over the security of future supplies, or, on the other hand, an anticipation by investors of a rise in the value of crude oil once refining capacity picks up again.

As for the global usefulness of the Hirsch conclusions, as of 2004, the US share of global oil consumption was about 26%, while its share of world population was only 4.3%; Europe used 11% of global oil while having about 6.8% of world population. An average car in Germany uses about 8.1 liter per 100 km; the US consumption 16.2 L. In US terms, 1 gallon delivers 44 miles in Germany but only 22 in the United States.

So far a part of the changes ultimately requested by Hirsch for the US have been already implemented in Europe (and *cum grano salis* in Asia). The difference had been much smaller at the start of the 70s. Europe adapted more after the various oil shocks and enhanced the changes by introducing much higher taxes on gasoline. The differences now are not only a lack of energy saving technologies, in car building and usage, and passive insulation of buildings in the US. The traditional significant differences in the setup and density of settlements, share of suburbs, use of public transport and consumer behavior have been widening. Taking this into account, a peak oil shock as outlined by Hirsch will have a much more severe outcome in the US compared to other parts of the world, especially Europe.

Mitigation of Peak Oil

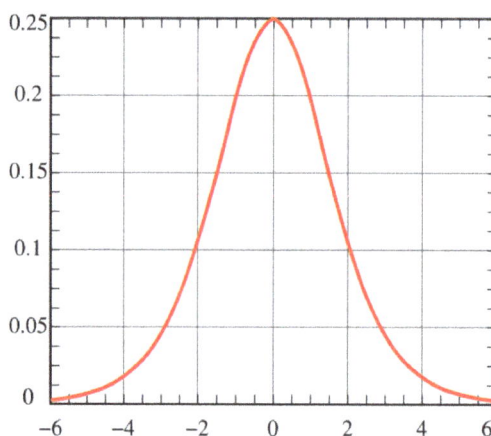

The standard Hubbert curve, plotting crude oil production of a region over time.

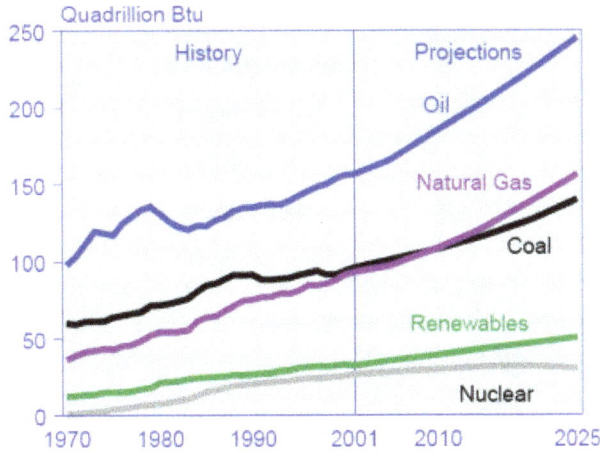

World energy consumption, 1970-2025. *Source: International Energy Outlook 2004.*

Part of a series about
Sustainable energy

Energy conservation
• Cogeneration
• Efficient energy use
• Green building
• Heat pump
• Low-carbon power
• Microgeneration
• Passive solar building design
Renewable energy
• Anaerobic digestion
• Geothermal
• Hydroelectricity
• Solar
• Tidal
• Wind

Sustainable transport
• Carbon-neutral fuel
• Electric vehicle
• Fossil-fuel phase-out
• Green vehicle
• Plug-in hybrid
Sustainable development portal
• Renewable energy portal
• Environment portal

The mitigation of peak oil is the attempt to delay the date and minimize the social and economic effects of peak oil by reducing the consumption of and reliance on petroleum. By reducing petroleum consumption, mitigation efforts seek to favorably change the shape of the Hubbert curve, which is the graph of real oil production over time predicted by Hubbert peak theory. The peak of this curve is known as peak oil, and by changing the shape of the curve, the timing of the peak in oil production is affected. An analysis by the author of the Hirsch report showed that while the shape of the oil production curve can be affected by mitigation efforts, mitigation efforts are also affected by the shape of Hubbert curve.

For the most part, mitigation involves fuel conservation, and the use of alternative and renewable energy sources. The development of unconventional oil resources can extend the use of petroleum, but does not reduce consumption.

Historically, world oil consumption data show that mitigation efforts during the 1973 and 1979 oil shocks lowered oil consumption, while general recessions since the 70s have had no effect on curbing the oil consumption until 2007.

Key questions for mitigation are the viability of methods, the roles of government and private sector and how early these solutions are implemented. The responses to such questions and steps taken towards mitigation may determine whether or not the lifestyle of a society can be maintained, and may affect the population capacity of the planet.

Alternative Energy

The most effective method of mitigating peak oil is to use renewable or alternative energy sources in place of petroleum.

Nuclear power, considered by some to be a viable alternative source, can be substituted for petroleum in some cases. China is preparing for the post-peak oil future by building pebble bed reactors configured to produce hydrogen fuel from the electrolysis of water. The use of nuclear power is often a highly contentious issue because of questions of the future availability of fuel and the dangerous nature of nuclear waste. Some current research projects are focused on neutron-free fusion power, in which hydrogen and boron are heated to over 1 billion degrees, though technical and economic barriers still exist.

In its October 2009 peak oil report, the Government-supported UK Energy Research Centre warned of the risk that 'rising oil prices will encourage the rapid development of carbon-intensive alternatives which will make it difficult or impossible to prevent dangerous climate change and stated that 'early investment in low-carbon alternatives to conventional oil is of considerable importance' in avoiding this scenario.

Iceland was the first country to suggest transitioning to 100% renewable energy, using hydrogen for vehicles and its fishing fleet, in 1998, but the actual progress has been very limited.

Transportation Fuel Use

Because most oil is consumed for transportation most mitigation discussions revolve around transportation issues.

Fuel Substitution

While there is some interchangeability, the alternative energy sources available tend to depend on whether the fuel is being used in static or mobile applications.

Biofuel

The use of biofuels, which are fuels derived from recently dead biological material, reduces dependence on petroleum and enhances energy security. Biofuels also play significant roles in the "food vs fuel" debate, mitigation of oil prices, and energy balance and efficiency. Ethanol is a biofuel produced from crops high in sugar (e.g., sugar cane, sweet sorghum) or starch, (corn/ maize). Biofuels can also be produced from plants that contain high amounts of vegetable oil, such as oil palm, soybean, algae, switchgrass, or jatropha. These oils can be burned directly in certain designs of diesel engines, or they can be chemically processed to produce fuels such as biodiesel. Wood and its byproducts can also be converted into biofuels such as woodgas, methanol or ethanol fuel. It is also possible to make cellulosic ethanol from non-edible plant parts, but this can be difficult to accomplish economically. Biofuels are most commonly used in vehicles, and in heating homes, and cooking. Biofuel industries are expanding in Europe, Asia and the Americas.

Several firms have successfully created petroleum products in the lab using either solid catalysts or genetically modified microorganisms. As of July, 2008, such firms are producing petroleum products in very small quantities, but hope to increase production over the next few years.

Static Installations

The substitution of oil with other fossil fuels is theoretically relatively easy when static installations are concerned, as in the case for electricity generation. Reserves of coal are substantial, and the technology to use it is well established. Increasing the use of coal, however, would lead to higher carbon emissions which is likely to be politically unacceptable in many countries due to the implications of global warming, although carbon capture and storage may provide a solution. Natural gas is another alternative, and combined cycle power generation using natural gas is the cleanest source of power available using fossil fuels, producing about 30% less carbon dioxide than burning petroleum and about 45% less than burning coal. The major difficulty in the use of natural gas is transportation and storage because of its low density. Natural gas pipelines are economical, but are impractical across oceans.

Mobile Applications

Due to its high energy density and ease of handling, oil has a unique role as a transportation fuel. There are, however, a number of possible alternatives. Among the biofuels the use of bioethanol and biodiesel is already established to some extent in some countries.

The use of hydrogen fuel is another alternative under development in various countries, alongside, hydrogen vehicles though hydrogen is actually an energy storage medium, not a primary energy source, and consequently the use of a non-petroleum source would be required to extract the hydrogen for use. Though hydrogen is quite clearly outclassed in areas of cost and efficiency by battery powered vehicles, there are applications where it would come in useful. Short haul ferries and very cold climates are two examples. Hydrogen fuel cells are about a third as efficient as batteries and double the efficiency of gasoline vehicles.

Electric vehicles powered by batteries are another alternative, and these have the advantage of having the highest well-to-wheels efficiency rate of any energy pathway and thus would allow much greater numbers of vehicles than any other methods. In addition, even if the electricity was sourced from coal-fired power plants, two advantages would remain: first it is cheaper to sequester carbon from a few thousand smokestacks than it is to retrofit hundreds of millions of vehicles, and second encouraging the use of electric vehicles allows a further pathway for scaling up of renewable energy sources.

Currently the cost of batteries capable powering electric vehicles for a 300-mile (480 km) range (comparable to the range of many gasoline vehicles) is prohibitively high, though producing batteries for plug-in hybrids with a 40-mile (64 km) range could be done with current technology and current pricing models within the reach of the average person. A plug-in hybrid with a 40-mile (64 km) range would have the advantage that it uses no gasoline or diesel at all for the first 40 miles (a distance coving 80% of all vehicle commutes).

Unfortunately there are currently no production models of plug-in hybrids or alternative fuel vehicles (other than flex fuel) available from big manufacturers, though both Toyota and General Motors have promised versions around 2010. Fully electric vehicles are available from Tesla Motors for their high priced sports car and also a small city vehicle from Th!nk in Norway, in limited production runs in Norway and the UK.

Alternative Aviation Fuel

The Airbus A380 flew on alternative fuel for the first time on 1 February 2008. Boeing also plans to use alternative fuel on the 747. Because some biofuels such as ethanol contains less energy, more "tankstops" might be necessary for such planes.

The US Air Force is currently in the process of certifying its entire fleet to run on a 50/50 blend of synthetic fuel derived from the Fischer-Tropsch process and JP-8 jet fuel.

Conservation

When alternative fuels are not available, the development of more energy efficient vehicles becomes an important mitigant. Some ways of decreasing the oil used in transportation include increasing the use of bicycles, public transport, carpooling, electric vehicles, and diesel and hybrid vehicles with higher fuel efficiency.

More comprehensive mitigations include better land use planning through smart growth to reduce the need for private transportation, increased capacity and use of mass transit, vanpooling and carpooling, bus rapid transit, telecommuting, and human-powered transport from current levels. Rationing and driving bans are also forms of reducing private transportation. The higher oil prices of 2007 and 2008 caused United States drivers to begin driving less in 2007 and to a much greater extent in the first three months of 2008.

In order to deal with potential problems from peak oil, Colin Campbell has proposed the Rimini protocol, a plan which among other things would require countries to balance oil consumption with their current production.

Unconventional Oil

Unconventional oil is oil produced or extracted using techniques other than the traditional oil well method from sources such as oil sands, oil shale and the conversion of coal or natural gas to liquid hydrocarbons through processes such as Fischer-Tropsch synthesis. Currently, unconventional oil production is less efficient and has a larger environmental impact relative to conventional oil production. Compared to conventional oil, much more energy is required to extract oil from non-conventional sources, so increasing costs and carbon emissions. Technology, such as using steam injection in oil sands deposits, is being developed to increase the efficiency of unconventional oil production.

Synthetic fuel, created via coal liquefaction, requires no engine modifications for use in standard automobiles. As a byproduct of oil embargoes during Apartheid in South Africa, Sasol, using the Fischer-Tropsch process, developed relatively low-cost coal-based fuel. Currently, about 30% of South Africa's transport-fuel (mostly diesel) is produced from coal. With crude-oil prices above US$40 per barrel, this process is now cost-effective.

Masdar, an Experiment in Mitigation

One government which is moving forward with mitigation plans is the emirate of Abu Dhabi. The United Arab Emirates economy minister stated in 2007 that the UAE do not believe that relying on oil revenues is sustainable, and so are moving to diversify their economies. Besides allotting land for solar power plants and partnering with Massachusetts Institute of Technology to build an alternative energy research institute, a new city is being constructed 17 kilometres (11 mi) east-southeast of the city of Abu Dhabi, which will rely entirely on solar energy, with a sustainable, zero-carbon, zero-waste ecology. Known as Masdar (Arabic for *source*), the initiative is headed by the Abu Dhabi Future Energy Company (ADFEC) The project is estimated to take some 10 years to complete, with the first phase complete and habitable in 2009, and a goal of housing 50,000 people and 1,500 businesses. The city is intended to cover 6 square kilometres (1,500 acres), with no point further than 200 m from a solar powered personal rapid transit link, housing energy, science and technology communities, commercial areas, a university, and the headquarters of the Future Energy Company. By relying on sustainable energy sources, keeping cars out of the city, returning to older architectural conventions (such as reducing air conditioning costs with large tents and narrow spaces between buildings), using sewage to produce energy and create soil, taking advantage of all recycling opportunities (including for and from construction), and reusing gray water, Masdar is designed to be a city which will consume no oil.

Bioplastics

Another major factor in petroleum demand is the widespread use of petroleum products such as plastic. These could be partially replaced by bioplastics, which are derived from renewable plant feedstocks such as vegetable oil, corn starch, hemp plants, pea starch, or microbiota. They are used either as a direct replacement for traditional plastics or as blends with traditional plastics. The most common end use market is for packaging materials. Japan has also been a pioneer in bioplastics, incorporating them into electronics and automobiles.

US Government Debate Over Mitigation Strategies

Part of the current debate revolves around energy policy, and whether to shift funding to increasing energy conservation, fuel efficiency, or other energy sources like solar, wind, and nuclear power. At congressional peak oil hearings, Rep. Tom Udall argued that while rising oil prices would encourage alternatives (both on the supply and demand side), the costs and impacts of other issues involved with petroleum based personal transportation (such as pollution, the economic effects of global warming, security threats caused by sending vast amounts of money to the Middle East, and the costs of road maintenance) should also be taken into account. "Because the price of oil is artificially low, significant private investment in alternative technologies that provide a long-term payback does not exist. Until oil and its alternatives compete in a fair market, new technologies will not thrive."

In 2005, the Congressional Budget Office suggested that, "the federal government could more effectively increase the efficiency of the nation's automotive fleet by raising gasoline taxes, imposing user fees on the purchase of low-mileage-per-gallon vehicles, or both." This would give automakers more incentive to research alternative fuel technology and increased efficiency (through lighter vehicles, better aerodynamics, and less wasted energy).

Hans-Holger Rogner, a section head at the IAEA, warned in 1997 that the level of incentive required for market driven research and development will actually rise. Because production costs are not expected to decrease and because of the continued emphasis companies give to short-term profits, "a regional breakdown for 11 world regions indicates that neither hydrocarbon resource availability nor costs are likely to become forces that automatically would help wean the global energy system from the use of fossil fuel during the next century."

The problems of privately funded research and development are not unique to peak oil mitigation. Bronwyn H. Hall, graduate economics professor at the Haas School of Business, points out that, "even if problems associated with incomplete appropriability of the returns to R&D are solved using intellectual property protection, subsidies, or tax incentives, it may still be difficult or costly to finance R&D using capital from sources external to the firm or entrepreneur. That is, there is often a wedge, sometimes large, between the rate of return required by an entrepreneur investing his own funds and that required by external investors." The severity of the problem for energy is echoed in the International Energy Agency's latest report

In the US, transportation by car is guided more by the government than by an invisible hand. Roads and the interstate highway system were built by local, state and federal governments and paid for by income taxes, property taxes, fuel taxes, and tolls. The Strategic Petroleum Reserve is designed to offset market imbalances. Municipal parking is frequently subsidized. Emission standards regulate pollution by cars. US fuel economy standards exist but are not high enough to have effect. There is also a gas guzzler tax of limited scope. The United States offers tax credits for certain vehicles and these frequently are hybrids or compressed natural gas cars (Energy Policy Act of 2005).

In order to be profitable, many alternatives to oil require the price of oil to remain above some level. Investors in these alternatives must gamble with the limited data on oil reserves available. This imperfect information can lead to a market failure caused by a move by nature. One explanation for this is Hotelling's rule for non-renewable resources. Even with perfect information the price of oil correlates with spare capacity and spare capacity does not warn of a peak. For example, in the early 1960s (10 years before oil production peaked in the United States), there was enough spare capacity in US production that Hubbert's predicted peak of 1966-1971 was "at the very least completely unrealistic to most people," preventing the necessary steps being taken to mitigate the situation. The absence of accurate information about spare production capacity exacerbates the current situation.

Lester Brown believes this problem might be solved by the government establishing a price floor for oil. A tax shift raising gas taxes is the same idea. Opponents of such a price floor argue that the markets would distrust the government's ability to keep the policy when oil prices are low.

In 2007, a Pentagon Report, "Space-Based Solar Power: An Opportunity for Strategic Security" proposed Space-Based Solar Power as a macro solution to peak oil and fossil fuel depletion. Recently a proposal for US leadership in SBSP won the SECDEF D3 competition. Engineer Keith Henson discussed the scale in "Dollar a Gallon Gasoline". Mike Snead has recently assessed prospects for US fossil fuels and SSP in "US fossil fuel energy insecurity and space solar power". Snead and Henso recently put out a video.

Implications of an Unmitigated World Peak

According to the Hirsch report prepared for the U.S. Department of Energy in 2005, a global decline in oil production would have serious social and economic implications without due preparation. Initially, an unmitigated peak in oil production would manifest itself as rapidly escalating prices and a worldwide energy crisis. While past oil shortages stemmed from a temporary insufficiency of supply, crossing Hubbert's Peak means that the production of oil continues to decline, so demand must be reduced to meet supply. If alternatives or conservation (orderly demand destruction) are not forthcoming, then disorderly demand destruction will occur, with the possible effect that the many products and services produced with oil become scarcer, leading to lower living standards.

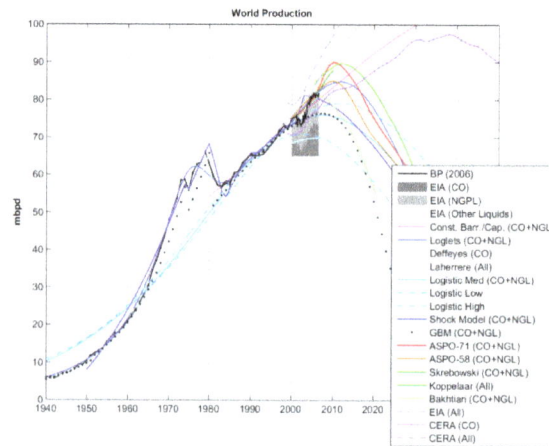

Oil depletion scenarios

- Air travel, using roughly 7% of world oil consumption, would be one of the affected services. The energy density of hydrocarbons and the power density of a jet engine are so necessary for aviation that hydrocarbon fuels are nearly impossible to replace with electricity, to an extent beyond any other common mode of transport.

- A US Army Corps of Engineers report on the military's energy options states

  "The Army and the nation's heavy use of oil and natural gas is not well coordinated with either the nation's or the Earth's resources and upcoming availability."

- Shipping costs

  "On average, a one percent increase in fuel prices leads to a 0.4% increase in total freight rates. Using this rule of thumb, the recent doubling in oil prices has raised averaged freight rates by almost 40%."

Shipping costs are particularly relevant to a country like Japan that has greater food miles.

- Increasing cost of oil for importing countries ultimately reduces those countries' purchase of non-oil goods abroad. The Federal Reserve Bank of San Francisco discusses oil and the US balance of trade:

  "Oil prices have almost quadrupled since the beginning of 2002. For an oil-importing country like the U.S., this has substantially increased the cost of petroleum imports. International trade data suggest that this increase has exacerbated the deterioration of the U.S. trade deficit, especially since the second half of 2004."

US indications of economic volatility have manifested themselves in the largest increase in inflation rates in 15 years (Sept. 2005), due mostly to higher energy costs.

- Significant oil producing countries will have a national purchasing advantage over similar countries with no oil to sell. This can result in larger militaries for oil producers or inflation of the price of whatever commodities they purchase. Saudi Arabia purchased US$40 billion worth of arms from the US between 1990 and 2000.

- The United States averaged 464 US gallons (1,760 L) of gas per person in 2004. Therefore, increased gasoline cost will likely make gas reducing alternatives increasingly necessary and common for lower income US residents.

Those who feel that much more drastic imminent social and cultural changes will occur from oil shortages are known as doomers.

References

- Greenwood, N. N.; & Earnshaw, A. (1997). Chemistry of the Elements (2nd Edn.), Oxford:Butterworth-Heinemann. ISBN 0-7506-3365-4.

- Deffeyes, Kenneth S (2002). Hubbert's Peak: The Impending World Oil Shortage. Princeton University Press. ISBN 0-691-09086-6.

- Madureira, Nuno Luis (2014). Key Concepts in Energy. London: Springer International Publishing. pp. 125–6. doi:10.1007/978-3-319-04978-6_6. ISBN 978-3-319-04977-9.

- P. Crabbè, North Atlantic Treaty Organization. Scientific Affairs Division (2000). "Implementing ecological integrity: restoring regional and global environmental and human health". Springer. p.411. ISBN 0-7923-6351-5

- Kunstler, James Howard (1994). Geography of Nowhere: The Rise And Decline of America's Man-Made Landscape. New York: Simon & Schuster. ISBN 0-671-88825-0

- Hirsch, Robert L.; et al. (2005). "PEAKING OF WORLD OIL PRODUCTION: IMPACTS, MITIGATION, & RISK MANAGEMENT" (PDF). US Department of Energy: 1–91. Retrieved 14 January 2016.

- Tokic, Damir (October 2015). "The 2014 oil bust: Causes and consequences". Energy Policy. 85: 162–169. doi:10.1016/j.enpol.2015.06.005. Retrieved 12 March 2016.

- Caruana, Jaime (February 5, 2016). "Credit, commodities and currencies" (PDF). Bank of International Settlements. Retrieved 12 March 2016.

- "Short - Term Energy Outlook March 2016 1 March 2016 Short - Term Energy Outlook (STEO)" (PDF). U.S. Energy Information Administration. Retrieved 12 March 2016.

- Baumeister and Kilian. "Understanding the Decline in the Price of Oil since June 2014". Social Science Research Network. CFS Working Paper No. 501. Retrieved 12 March 2016.

- Shilling, A. Gary (20 August 2015). "A Funny Thing Happened on the Way to $80 Oil". Bloomberg News. Bloomberg L.P. Retrieved 6 November 2015.

- Brady, Jeff (December 18, 2014). "Citing Health, Environment Concerns, New York Moves To Ban Fracking". National Public Radio. Retrieved 6 January 2015.

- McSpadden, Kevin. "China Has Become the World's Biggest Crude Oil Importer for the First Time". TIME. Time Inc. Retrieved 16 August 2015.

- Davis, Bob. "IMF Warns of Slower China Growth Unless Beijing Speeds Up Reforms". The Wall Street Journal. News Corp. Retrieved 16 August 2015.

- Schenk, C.J.; et al. "An Estimate of Recoverable Heavy Oil Resources of the Orinoco Oil Belt, Venezuela" (PDF). United States Geological Survey. Retrieved 16 November 2015.

- King D., Murray J. (2012). "Climate policy: Oil's tipping point has passed.". Nature. 481 (7382): 433–435. doi:10.1038/481433a. Retrieved 23 September 2015.

- Auzanneau, Matthieu (17 March 2014). "Nouvelle chute en 2013 de la production de brut des " majors ", désormais contraintes à désinvestir". Le Monde. Retrieved 26 April 2014.

Permissions

Index

www.ingramcontent.com/pod-product-compliance
Lightning Source LLC
Chambersburg PA
CBHW061250190326
41458CB00011B/3634